Steel Corrosion-Induced Concrete Cracking

Steel Corrosion-Induced Concrete Cracking

Yuxi Zhao
Zhejiang University, Hangzhou, China

Weiliang Jin
Zhejiang University, Hangzhou, China

AMSTERDAM • BOSTON • HEIDELBERG • LONDON
NEW YORK • OXFORD • PARIS • SAN DIEGO
SAN FRANCISCO • SINGAPORE • SYDNEY • TOKYO
Butterworth-Heinemann is an imprint of Elsevier

Butterworth-Heinemann is an imprint of Elsevier
The Boulevard, Langford Lane, Kidlington, Oxford OX5 1GB, UK
50 Hampshire Street, 5th Floor, Cambridge, MA 02139, USA

Notices
Knowledge and best practice in this field are constantly changing. As new research and experience broaden our understanding, changes in research methods, professional practices, or medical treatment may become necessary.

Practitioners and researchers must always rely on their own experience and knowledge in evaluating and using any information, methods, compounds, or experiments described herein. In using such information or methods they should be mindful of their own safety and the safety of others, including parties for whom they have a professional responsibility.

To the fullest extent of the law, neither the Publisher nor the authors, contributors, or editors, assume any liability for any injury and/or damage to persons or property as a matter of products liability, negligence or otherwise, or from any use or operation of any methods, products, instructions, or ideas contained in the material herein.

ISBN: 978-0-12-809197-5

British Library Cataloguing-in-Publication Data
A catalogue record for this book is available from the British Library.

Library of Congress Cataloging-in-Publication Data
A catalog record for this book is available from the Library of Congress.

Publisher: Joe Hayton
Acquisition Editor: Simon Tian
Editorial Project Manager: Naomi Robertson
Production Project Manager: Lisa Jones
Cover Designer: Greg Harris

Typeset by MPS Limited, Chennai, India

Contents

Foreword

Service life of unreinforced concrete structures can be extremely long. Many historical structures from the Roman period demonstrate this fact convincingly, such as the Pantheon in Rome, which is built of high-strength concrete, normal concrete, and a cupola of extremely lightweight concrete, or Roman bridges all over Europe. They have survived thousands of years under the effects of natural exposure, wars, and earthquakes without protection or maintenance.

If steel reinforcement is embedded in concrete, then the steel is protected initially by a surface passivation layer in the aqueous pore solution because of its high pH value. This passivation layer, however, is not stable in aging concrete. Concrete is a porous material. At first, the pore space is partially filled with water and is in equilibrium with high relative humidity. However, the surface of reinforced concrete structures is often exposed to an atmosphere with much lower humidity. As a consequence, a long-lasting drying process begins as soon as the formwork is removed. As the pore space is partially emptied, gases diluted in the surrounding air such as CO_2 will slowly migrate into the pore space and react with the young cement-based matrix. First, a thin layer of carbonated hydration products is formed near the surface. When the growing thickness of this carbonated layer reaches the cover thickness, the pH value of the pore liquid in contact with the steel reinforcement is lowered and the protective passivation layer is eventually destroyed. Hence, corrosion is initiated.

If the surface of concrete is temporarily in contact with an aqueous chloride solution such as sea water or water containing dissolved deicing salt, then chloride ions can penetrate into the pore space of the concrete. Fortunately, the chloride ions are filtered out of the salt solution and they are enriched close to the surface; however, the clean water may penetrate deep into the pore space by capillary action. Later, dissolved chloride ions migrate deeper into the nanoporous material by slow diffusion. Whenever the chloride concentration near the steel reinforcement reaches a critical value, the protective passivation layer may be locally destroyed and corrosion of steel begins.

Until now, durability design had not reached the sophisticated level of structural design. Usually one considers different stages, for instance, corrosion initiation, crack formation, and spalling. After initiation, the rate of corrosion may be rather low. But after the first cracks are generated, corrosion will be significantly accelerated. The authors of this book primarily describe the formation of corrosion products and the time-dependent pressure built up by the voluminous corrosion products. These are essential processes that may serve as indications of the end of service life of reinforced concrete structures.

The mechanisms of steel corrosion in concrete are briefly described. The following chapters describe in great detail the formation of corrosion products at the interface between steel and concrete. In particular, the gradual filling of the pores of hardened cement paste and the gradual filling of cracks with corrosion products are considered. Finally, a model describing concrete cracking due to the formation of corrosion products is presented. This approach and the results obtained by the two authors of this book will be very helpful for more realistic prediction of the service life of reinforced concrete structures.

In this book, the complex processes that can be observed during the critical period of aging concrete between corrosion initiation and crack formation in the concrete cover and then, finally, spalling of the layer near the surface are described in great detail for the first time. A better understanding of these complex processes will contribute to the development of more reliable prediction of service life and of more efficient methods for protective interventions. Therefore, the service life of reinforced concrete structures can be extended systematically and the cost of repair measures can be substantially reduced. This is of particular importance for countries like China, in which a nationwide infrastructure is rapidly being built. In no other country is more concrete being produced at this moment.

If modern construction does not become more sustainable, and if the service life of reinforced concrete structures is not significantly extended, then further development will be slowed because of the enormous costs for repair and maintenance of the existing infrastructure. This book may contribute a great deal to finding realistic and sustainable solutions to this worldwide problem.

This volume deserves wide distribution and will hopefully be studied in great detail by scientists and practitioners. With this book, the actual situation concerning durability and service life in construction can be improved substantially worldwide.

Folker H. Wittmann

Preface

Since the mid 1970s, a number of durability-related problems have emerged and stimulated research of the key factors that relate to the durability problems of concrete structures. Reinforcing steel corrosion is one of the major reasons for deterioration of reinforced concrete structures. Steel corrosion in concrete will induce sectional loss of steel bar, degradation of bond stress between steel and concrete, and cracks parallel to longitudinal bars. Field studies have suggested that cracking and spalling are most concerning to asset owners. Structural collapse of reinforced concrete structures due to steel corrosion is rare; cracking, rust staining, and spalling of the concrete cover usually appear well before a reinforced concrete structure is at risk. Therefore, the cracking of the concrete cover induced by steel corrosion is important and is usually defined as the serviceability limit state.

The authors have focused on this research field since 1998, and have carefully investigated the origin, mechanism, and development of corrosion-induced cracking in concrete. Considering the importance of this topic, the authors summarize the related research achievements obtained and share with other researchers and engineers who are interested in this field. This book concentrates on the concrete cracking process induced by steel corrosion. After the background introduction and literature review in chapter "Introduction," the mechanisms of steel corrosion in concrete are introduced in chapter "Steel Corrosion in Concrete." The composition, expansion coefficient, and elastic modulus of steel corrosion are carefully investigated in chapter "The Expansion Coefficients and Modulus of Steel Corrosion Products," considering the importance of the properties of steel corrosion in concrete cracking models. With these parameters of steel corrosion, the damage analysis is applied to analyze the corrosion-induced concrete cracking process in chapter "Damage Analysis and Cracking Model of Reinforced Concrete Structures With Rebar Corrosion," and the critical thickness of the rust layer at the moment of surface cracking of concrete cover is studied in chapter "Mill Scale and Corrosion Layer at Concrete Surface Cracking." In chapter "Rust Distribution in Corrosion-Induced Cracking Concrete," the authors investigate the rust distribution in the corrosion-induced cracks and find that the rust did not fill the corrosion-induced cracks in the concrete cover before concrete surface cracking. A Gaussian function is proposed to describe the nonuniform spatial distribution of corrosion products in chapter

"Nonuniform Distribution of Rust Layer Around Steel Bar in Concrete." The shape of the corrosion-induced cracks in the concrete cover is observed in chapter "Crack Shape of Corrosion-Induced Cracking in the Concrete Cover," and a linear model was proposed to describe the variation in the total circumferential crack width along the radial direction in the concrete cover. Rust distribution at the steel–concrete interface is presented in chapter "Development of Corrosion Products-Filled Paste at the Steel–Concrete Interface"; and the penetration of corrosion products into the porous zone of concrete and formation of a corrosion layer at the steel–concrete interface process simultaneously. Finally, in chapter "Steel Corrosion-Induced Concrete Cracking Model," an improved corrosion-induced cracking model is proposed, which considers the corrosion layer accumulation and corrosion products filling occurring simultaneously in concrete. The time from corrosion initiation to concrete surface cracking is discussed. The need for more research regarding the corrosion-induced cracking model is also discussed in this book.

The authors hope this book is useful for researchers interested in the durability of concrete and concrete structure fields, for industry engineers who pay attention to the deterioration mechanisms and the life cycle of reinforced concrete structures, and for graduate students whose research topics include corrosion-induced deterioration of reinforced concrete structures.

Acknowledgments

Continuous financial support from the National Science Foundation of China (NSFC) through grants 50538070, 50920105806, 50808157, and 51278460 is gratefully acknowledged. Without this financial support, it would be not possible to perform all the experimental work, develop the analytical models, and further the research in this area.

The authors also express sincere gratitude to the graduate students who contributed their talent, intelligence, effort, and hard work to this research. They are Haiyang Ren, Jiang Yu, Bingyan Hu, Hong Dai, Yingyao Wu, Hangjie Ding, Jianfeng Dong, and Xiaowen Zhang. Particular thanks are expressed to Hangjie Ding, Jianfeng Dong, and Xiaowen Zhang, who also helped to edit the contents, improve the format, and modify some figures in this book.

We appreciate the comments and encouragement from Professor Folker H. Wittmann, his comments improve the quality of the book, he also kindly wrote the *Foreword*. Many thanks to the reviewers of this book for their time. Their helpful suggestions have strengthened the book considerably. Our thanks also go to the publisher and the editor Fanjie Wu, who patiently replied to all our questions and helped us as much as possible.

We apologize for naming such a small number of the many people who helped us and with whom we had the pleasure of working. Sincere thanks to all of them for their willingness to share their knowledge with us and for encouraging us to go further and further in this area.

List of Figures

List of Tables

Chapter 1

Introduction

Chapter Outline

1.1 BACKGROUND

Since Portland cement was invented in 1824, it has been the most widely used construction material in the world. Until the mid-1970s, concrete was viewed as being maintenance-free and having a practically unlimited life. However, since then, a number of durability-related problems, such as corrosion of reinforcement, alkali-silica reactions, and sulfate attacks, have emerged and stimulated research investigating the key factors that relate to concrete durability in developed and developing countries.

Among all durability problems, reinforcement corrosion has been identified as the primary cause of deterioration in concrete structures [1]. The increased incidence of reinforcement corrosion is primarily due to two independent reasons. The first reason is the carbonation of the cover concrete and, thus, the loss of alkalinity. The second reason is the presence of chloride ions in sufficient amounts at the steel surface. Chloride ions are the primary cause of the initiation of reinforcement corrosion, and exposing concrete structures to environments rich in this ion significantly enhances the risk of structural degradation.

After the onset of corrosion, because the ferrous or ferric ions might form complexes, most of which occupy a larger volume than the original metal, the resulting increase in volume associated with the formation of corrosion products will exert outward expansive pressure on the surrounding concrete,

Steel Corrosion-Induced Concrete Cracking. DOI: http://dx.doi.org/10.1016/B978-0-12-809197-5.00001-3

eventually leading to the cracking of the concrete cover. These cracks, in turn, provide a path for a more rapid ingress of aggressive agents to the reinforcement, which can accelerate the steel corrosion process, thereby causing damage such as the delamination or spalling of the concrete cover. The subsequent reduction in the bond between the steel and concrete and the loss of the rebar cross-sectional area will, in time, adversely affect the serviceability and load capacity of the reinforced concrete elements. Therefore, in practice, many owners of concrete infrastructures use the appearance of visible cracks as an indication to execute appropriate interventions.

Corrosion-induced concrete cracking has received considerable attention from academics and practical engineers. The total service life of reinforced concrete elements subject to corrosion is commonly represented as a two-stage process consisting of an initiation period and a propagation period [2]. The former is defined as the time comprising the onset of reinforcement corrosion, and the latter is defined as the time to reach specified serviceability or ultimate limit states. The service life has traditionally been considered equal to the duration of the initiation period; consequently, much of the research has been focused on this area. However, this approach may be regarded as too conservative of a failure criterion because the end of the initiation phase only signifies the onset of corrosion, and the structure is yet to suffer any adverse effects that may inhibit its functional performance [3]. Over the past several years, greater emphasis has been placed on investigating the propagation phase. This approach enables the consideration of more realistic failure criteria to signify the end of the structural service life, such as the initial full-depth cracking of the cover, cracks exceeding a maximum allowable width, or even cover spalling and eventual structural collapse [4].

This chapter reviews a number of empirical and analytical models to describe the cracking process of concrete induced by steel corrosion. This book presents an investigation of the propagation period and focuses on corrosion-induced concrete in the cracking process. The contents of the whole book are also introduced in Section 1.4.

1.2 EMPIRICAL MODELS

A considerable number of experimental studies have been conducted regarding corrosion-induced concrete cracking. These have primarily focused on the following two aspects of the cracking process: (1) predicting steel corrosion at surface cracking and (2) linking the crack width on the surface of the concrete cover with steel corrosion.

1.2.1 Critical Steel Corrosion at Surface Cracking

Because of the importance of the concrete surface cracking moment, a number of researchers have conducted experimental work and developed empirical models. The tests were generally finished under accelerated conditions, most of which were through applying external currents. The initial cracking was marked by the appearance of the first visible crack (crack width less than 0.5 mm). The empirical models, as stated here, are primarily based on the regression analysis of the experimental data. In the following reviewed models, δ_{cr} (in μm) is the critical corrosion penetration (ie, the steel corrosion penetration at the concrete cover surface cracking moment), ρ_{cr} (in %) is the critical steel corrosion (ie, the steel cross-sectional loss area as the percentage of the original rebar cross-sectional area at the concrete cover surface cracking moment), t (in years) is the time, C (in mm) is the concrete cover thickness, d (in mm) is the steel bar diameter, and i_{corr} (in μA/cm^2) is the corrosion current density.

Andrade et al. [5] conducted accelerated corrosion experiments, and external currents of 10 μA/cm^2 and 100 μA/cm^2 were applied for different specimens. The results showed that regardless of i_{corr}, visible cracks (0.05–0.1 mm) were generated after a negligible bar cross-section loss of 10–20 μm. The critical corrosion penetration δ_{cr} was measured as follows:

$$\delta_{cr} = 11.6 \cdot i_{corr} \cdot t \tag{1.1}$$

Rodriguez et al. [6] also conducted accelerated corrosion experiments by applying the same current density as Andrade [5] and observed that the critical corrosion penetration δ_{cr} primarily depended on the cover-to-steel bar diameter ratio, C/d, and the concrete tensile strength, f_t, (in MPa). The proposed model, currently used by DuraCrete [7], is as follows:

$$\delta_{cr} = 83.8 + 7.4C/d + 22.6f_t \tag{1.2}$$

By conducting accelerated corrosion tests similar to those of Andrade [5] and Rodriguez [6], Alonso et al. [3] further investigated the effect of the corrosion current density, i_{corr}, by applying external currents of 3 μA/cm^2, 10 μA/cm^2, and 100 μA/cm^2. The results showed that the critical corrosion penetration, δ_{cr}, was only slightly affected by the corrosion current density, i_{corr}, and the following model was proposed based on the results:

$$\delta_{cr} = 7.53 + 9.32C/d \tag{1.3}$$

Zhang [8] conducted similar accelerated corrosion tests considering the effect of stirrups and the location and interaction of the steel bars. As a result, the following model was proposed:

$$\rho_{cr} = 1.43k_1k_2k_3(2C/d)^{1.287}[1 + (E_c \times 10^{-4} - 2.57)^2]\,f_c^{0.656} \tag{1.4}$$

where k_1 is the location coefficient of the bar and $k_1 = 1.26$ for bars in the internal location. k_2 is the interaction coefficient between bars, and $k_2 = 0.65$, 0.76, 0.81, and 1 when the spaces between bars are d, 1.5d, 2d, and larger than 5d, respectively. k_3 is the stirrups coefficient, and $k_3 = 1.25$, 1.15, 1.05, and 1 when the spaces between stirrups are 100 mm, 150 mm,

200 mm, and larger than 200 mm, respectively. E_c and f_c are the elastic modulus and compressive strength of the concrete, respectively.

Oh et al. [9] measured the concrete strain to determine the cracking state of the concrete cover. Once the concrete strain reached the ultimate value, the concrete cover was thought to be cracked. The following model was proposed to calculate steel section loss at surface cracking:

$$\rho_{cr} = 0.0018C^{2.07} \tag{1.5}$$

Webster [10] conducted regression analysis on 50 sets of experimental data from other researchers and proposed the following simple model for an approximate estimation of the critical corrosion depth:

$$\delta_{cr} = 1.25C \tag{1.6}$$

For some exciting empirical models about critical steel corrosion, their predictable performance, and a comparison of these models are discussed in Section 1.2.3.

1.2.2 Crack Width at the Concrete Surface

The initial cracks provide a path for rapid ingress of aggressive agents to the reinforcement, which can accelerate the steel corrosion process and cause damage such as delamination or spalling of the concrete cover. In practice, many owners of concrete infrastructures use the appearance of visible cracks as an indication to execute an appropriate intervention. Therefore, the corrosion-induced crack width is used as an important parameter to assess the serviceability and durability of the reinforced concrete structures.

Alonso et al. [3] observed that 15–50 μm of steel corrosion penetration can generate a 0.05–0.1mm crack width at the cover surface, whereas a 0.2–0.3mm crack width needs 50–200 μm of steel corrosion penetration.

Considering the location of steel bars in beams, Rodriguez et al. [6] proposed the following model to calculate the crack width on the concrete surface, W_s (in mm):

$$W_s = 0.05 + \beta[\delta - (83.8 + 7.4C/d + 22.6f_t)] \tag{1.7}$$

where δ (in μm) is steel radial loss induced by corrosion, $\beta = 0.0100$ for the steel bars at the top of the beam, and $\beta = 0.0125$ for the steel bars at the bottom.

Wang [11] proposed the following model:

$$\delta = 234.5W_s + 17.5 \tag{1.8}$$

Vu et al. [12] defined the *concrete quality*, which is the ratio between the cover thickness, C, and the water-to-cement ratio, w/c, or $C/(w/c)$, and observed that there is a nonlinear line of best fit before the crack width limit is reached. These researchers proposed the following model:

$$t = A\left(\frac{C}{w/c}\right)^B \tag{1.9}$$

where A and B are constants. Additionally, Vu [12] proposed a model to predict the time to cracking when extrapolating the accelerated corrosion test results to the behavior of real reinforced concrete structures. The time to crack initiation and crack propagation for real reinforced concrete structures, $t_{(real)}$, with any value of corrosion rate, $i_{corr(real)}$, is:

$$t_{(real)} = k_R \frac{i_{corr(exp)}}{i_{corr(real)}} t_{(exp)} \tag{1.10}$$

where $t_{(exp)}$ and $i_{corr(exp)}$ are the time to crack initiation and cracking propagation and the corrosion rate for the acceleration corrosion tests, respectively. k_R is the rate of loading correction factor, which can be calculated as:

$$k_R \approx 0.95\left[\exp\left(-\frac{0.3i_{corr(exp)}}{i_{corr(real)}}\right) - \frac{i_{corr(exp)}}{2500i_{corr(real)}} + 0.3\right] \tag{1.11}$$

Mullard et al. [13] investigated reinforced concrete slabs to study cover cracking propagation considering the effect of the concrete strength, cover thickness, reinforcement bar diameter, and concrete confinement. Citing the results from Vu [12], the following new model was proposed:

$$t_{(exp)} = \frac{k_R(W_s - 0.05)}{0.0008k_c\exp[-1.7C/(d\cdot f_t)]}\left(\frac{0.0114}{i_{corr}}\right) \tag{1.12}$$

where k_R has been shown in Eq. (1.11) and k_c is the confinement factor that represents an increase in crack propagation due to the lack of concrete confinement around the external reinforcing bar. $k_c = 1$ for a steel bar in the internal location. Eq. (1.12) is suitable for $0.1 \leq C/(d\cdot f_t) \leq 1$ and $W_s \leq 1.0$ mm.

Vidal et al. [14] investigated two beams naturally corroded in a saline environment and subjected to wetting–drying cycles over periods of 14 and 17 years. The results show that the cover/diameter ratio (C/d) and the bar diameter, d, seem to have no effect on the crack width evolution as a function of steel cross-section loss, and the following model was proposed:

$$W_s = 0.0575(\Delta A_s - \Delta A_{s0}) \tag{1.13}$$

where ΔA_s (in mm^2) is the steel loss of the cross-section and ΔA_{s0} (in mm^2) is the steel cross-section loss at the initial cracking moment. These factors can be calculated as:

$$\Delta A_s = \frac{\pi}{4}(2\alpha\delta d - \alpha^2\delta^2) \tag{1.14}$$

$$\Delta A_{s0} = A_s\left[1 - \left(1 - 0.001\alpha\frac{7.53 + 9.32C/d}{d}\right)^2\right] \tag{1.15}$$

where α is the pit concentration factor and $\alpha = 2$ for the even corrosion case. A_s is the original steel cross-section. When predicting steel corrosion at the initial cracking, Eq. (1.15) was used [3].

Zhang et al. [15] further investigated two beams exposed to a chloride environment for 14 and 23 years without any artificial accelerating measures. Based on other results [3,14], the following new model was proposed for predicting the crack width of reinforced concrete in a natural environment:

$$W_s = 0.1916\Delta A_s + 0.164 \tag{1.16}$$

These models are compared in Section 1.2.3.

1.2.3 Discussion on the Empirical Models

Based on the discussion in Sections 1.2.1 and 1.2.2, it can be found that most empirical models relate the critical corrosion penetration, δ_{cr} (or critical steel corrosion ρ_{cr}) and the surface crack width, W_s, to geometrical parameters such as the cover thickness, C, the steel bar diameter, d.

To assess the predictive performance of these models, Fig. 1.1 was drawn to compare steel corrosion on surface cracks calculated by empirical models [3,6,8,9,13,14] with corrosion observed in accelerated corrosion tests [3,5,6,8–10]. Fig. 1.2 compares the surface crack width propagation between model-predicted results [6,11,13–15] and the experimental results from Vu's work [12].

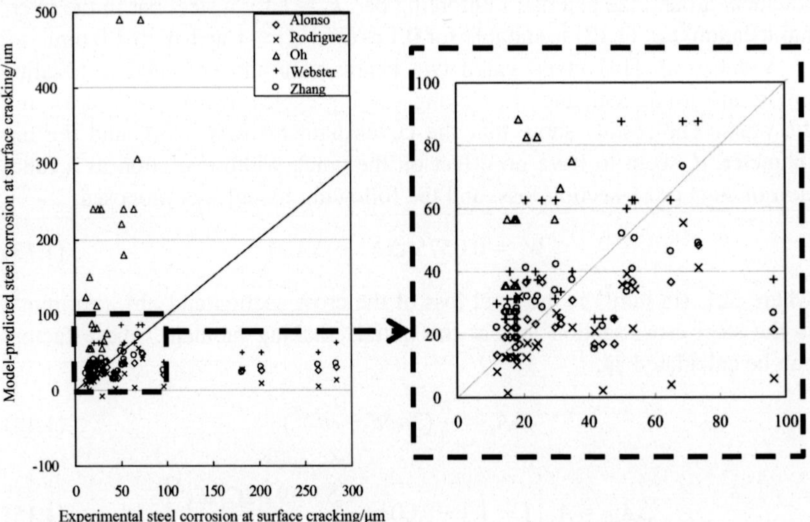

FIGURE 1.1 Comparison of steel corrosion at concrete surface cracking between the empirical model-predicted results and the experimental results.

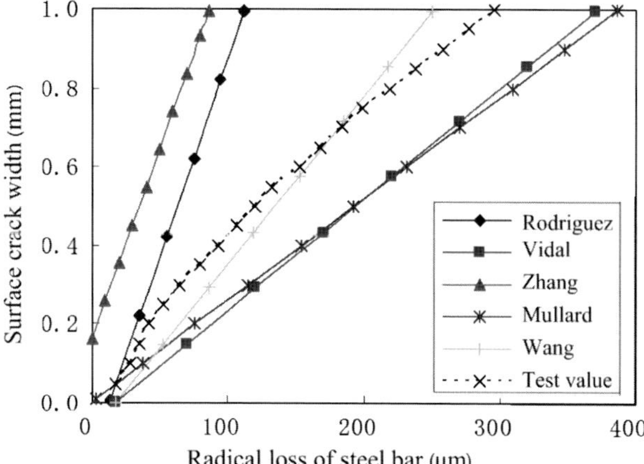

FIGURE 1.2 Comparison of concrete surface crack width propagation between the model-predicted results and the experimental results.

It can be clearly observed from Figs. 1.1 and 1.2 that none of the empirical models provides a good prediction of all of the experimental data. The empirical models are usually capable of adequately predicting the experimental values to which they were fitted. However, if the models are used to predict experimental data of other studies, then the agreement between the predicted and observed values is generally low. The existing empirical models are not universally valid, possibly because these empirical models usually cannot take into account all of the relative parameters. Furthermore, there are also some parameters, such as the "porous zone" around the parameter of the steel bar that accommodates corrosion products, that are not yet considered in the model. Therefore, analytical methods, which are able to consider more relevant parameters such as the geometrical dimensions, concrete properties, and steel corrosion, were used to study corrosion-induced cracking processes.

1.3 ANALYTICAL MODELS

1.3.1 Three-Stage Corrosion-Induced Cracking Model

The three-stage corrosion-induced cracking model has been widely accepted since the concepts of the three stages, namely, the corrosion products filling stage, the concrete cover stressing stage, and the concrete cover cracking stage (as shown in Fig. 1.3), were proposed by Liu and Weyers [16]. After steel depassivation, corrosion products are generated, which fill the porous voids around the steel–concrete interface. During this stage, the formation of corrosion products is assumed to not create extra stresses in the surrounding concrete, and the volume increase is compensated by the filling of the voids. Therefore, this stage is also regarded as the "rust-free expansion"

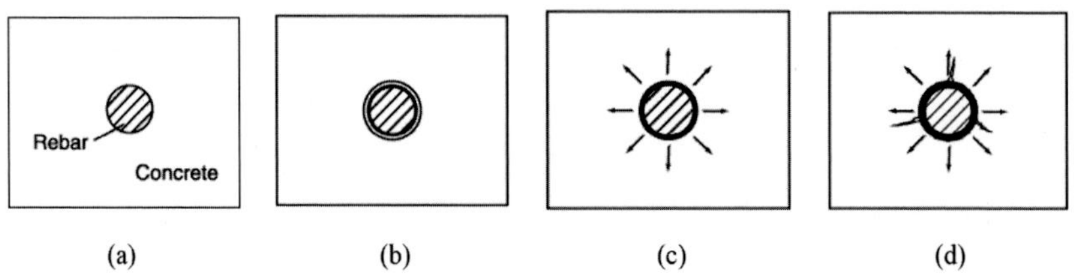

(a) (b) (c) (d)

FIGURE 1.3 Three-stage corrosion-induced cracking process. (a) Corrosion initiated. (b) Stage 1: filling. (c) Stage 2: stressing. (d) Stage 3: cracking.

stage [16]. As the corrosion products further increase, expansive pressure in the surrounding concrete is generated. This pressure increases with the corrosion products and creates extra stresses and strains in the concrete cover surrounding the reinforcing steel. When the internal tensile stresses exceed the tensile strength of the concrete cover, cracks will start in the concrete cover starting from the steel surface toward the concrete surface. During the process of the cracking propagation, the corrosion products will also fill the space within the cracks during this stage.

According to the three-stage corrosion-induced cracking model, the total radial loss of the steel bar when the concrete cover has cracks can be taken as approximately the summation of three components [16]. The first is the radial loss of the steel bar, δ_{pore}, produced during stage 1. The second is the radial loss of the steel bar, δ_{stress}, produced between the stress initiation in the concrete cover and the surface cracking of the concrete cover. The third is the radial loss of the steel bar, δ_{crack}, in which the corrosion products fill the space in the cracks in the concrete cover. Mathematically, this can be expressed as follows:

$$\delta_{cr} = \delta_{pore} + \delta_{stress} + \delta_{crack} \tag{1.17}$$

1.3.2 Corrosion Products Filling Stage

During the rust-free expansion stage, the concept of the so-called porous zone is assumed to accommodate the steel corrosion, which is generated freely in this stage without stressing the surroundings. However, its present form is highly questionable because it is not certain whether any reliable value can be associated with it [17].

Previous studies [18–25] have observed the area of concrete around the rebar penetrated by corrosion products. Wong et al. [23] observed this area (Fig. 1.4) and called it corrosion products-filled paste (CP), which was also called the corrosion-accommodating region by Michel [24]. CP is used to describe the area of concrete around the rebar penetrated by corrosion products in this chapter. The existence of CP has been verified for reinforced concrete members that had deteriorated in the natural environment [18–20],

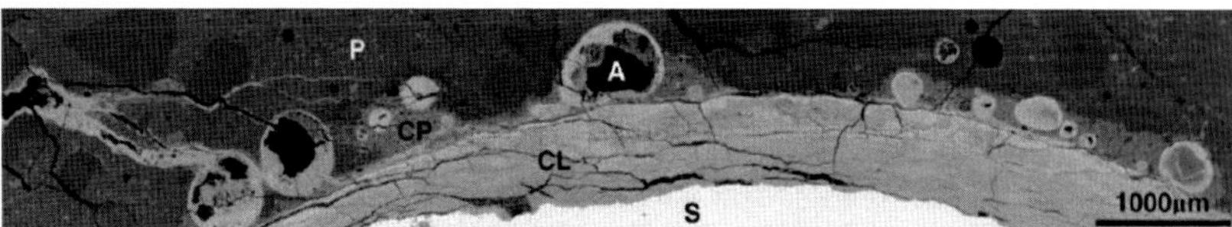

FIGURE 1.4 BSE images showing accumulation of corrosion products at the steel–concrete interface (S, steel; CL, corrosion layer; CP, corrosion products-filled paste; P, unaltered paste; A, air void) [23].

TABLE 1.1 Experimental Study of Corrosion Products-Filled Paste

Reference	Corrosion Method	Stage 1: Corrosion Products-Filled Paste
Caré et al. [21]	External current	The existence of corrosion products-filled paste was verified. It was pointed out that the corrosion products-filled paste was not in the whole area around the steel. When the rust filled the pores, it produced pressure on the paste layer but was not studied quantitatively.
Wong et al. [23]	Dry–wet cycle	The average depth of corrosion products-filled paste with different corrosion degrees is similar, approximately 100–200 μm, and is distributed on the side of serious steel corrosion.
Michel et al. [24]	External current	Corrosion products-filled paste increases along with the corrosion-induced cracking process, and the maximum value can reach 0.6 mm when the corrosion-induced crack occurs.
Zhao et al. [25]	External current	The corrosion products-filled paste was observed by SEM but not studied quantitatively.
Zhao et al. [26]	Dry–wet cycle	The penetration of corrosion products into the porous zone of concrete and the formation of a corrosion layer at the steel–concrete interface may occur simultaneously.
Zhao et al. [27]	External current	Proved that the penetration of corrosion products into the porous zone of concrete and the formation of a corrosion layer at the steel–concrete interface may occur simultaneously

during the artificial cyclic wet–dry tests [22,24,26], and in electrochemical corrosion conditions [21,24,25,27].

Table 1.1 presents experimental studies of CP. For example, Michel et al. [24] investigated the amount of corrosion products penetrating concrete and produced a semi-quantitative description of their changes during the electrochemical corrosion process. These results indicated that the depth of the CP increased with corrosion time, and the thickness reached 0.09−0.18 mm when the first crack occurred. In another study [23] the thickness of the CP was measured in samples with various degrees of corrosion, and the results showed that the CP was distributed primarily on the side of the concrete cover and was approximately 100−200 μm.

Scanning electron microscopy (SEM) and energy spectrum analysis (EDS) were used to study the corrosion-induced crack process of reinforced concrete specimens with accelerated corrosion. The results show that the penetration of corrosion products into the porous zone of concrete and the formation of a corrosion layer at the steel−concrete interface may occur simultaneously after the initiation of steel corrosion. Detailed information of these results is presented in chapters "Rust Distribution in Corrosion-Induced Cracking Concrete" and "Development of the Corrosion Products-Filled Paste at the Steel/Concrete Interface."

1.3.3 Concrete Cover Stressing and Cracking

Many investigations in the literature describe the corrosion-induced concrete cover stressing and cracking process, which is mainly related to the mechanical performance of a concrete cover subject to internal pressure. Table 1.2 gives a summary of the current models.

It can be observed that the previous model usually considered the concrete around the steel as a one-layer or two-layer thick-walled cylinder subjected to internal pressure due to the formation of corrosion products, as shown in Fig. 1.5a. In the double-layer thick-walled cylinder, the concrete cylinder can be divided into two coaxial cylinders. One is the inner cracked cylinder and the other is the outer intact cylinder, as shown in Fig. 1.5b. For example, Bazant [28] proposed a model to predict the time of concrete cover cracking caused by the corrosion of embedded reinforcing steel. The model considers the concrete around the steel as a thick-walled cylinder subjected to internal pressure due to the formation of corrosion products. Stresses in the cylinder are calculated using the solution provided by linear elasticity theory. Bhargava [29] proposed a double-layer thick-walled cylinder model to predict the time required for cover cracking and the weight loss of reinforcing steel. The model considers the residual strength of cracked concrete and the stiffness contribution from the combination of reinforcement and expansive corrosion products. However, the mechanical properties of the combination were assumed to be the same as the reinforcement in this model. It can been seen from Table 1.2 that, apart from the elasticity theory, the fracture mechanics and the damage mechanics are also introduced into some corrosion-induced concrete cracking analytical models.

It also can be determined that most existing models neglected the behavior of the rust layer between the steel bars or assumed the values of the rust parameters. However, it is clear that the steel corrosion not only causes the

TABLE 1.2 Corrosion-Induced Cracking Model

Reference	Method	Model	Residual Stiffness of Cracking Concrete	Rust Performance
Liu et al. [17]	Elastic mechanics	Thick-walled cylinder	Not considered	Not considered
Bazant [28]	Elastic mechanics	Thick-walled cylinder	Not considered	Not considered
Bhargava et al. [29]	Elastic mechanics	Double-layer thick-walled cylinder	Different elastic modulus in cracking and uncracking concrete	Not considered
Zhao et al. [30]	Elastic mechanics	Thick-walled cylinder	Not considered	Considered
Li et al. [31]	Fracture mechanics, elastic mechanics	Double-layer thick-walled cylinder	Smeared crack, considering the softening of cracking concrete	Not considered
Yu et al. [32]	Damage mechanics, elastic mechanics	Double-layer thick-walled cylinder	Smeared crack, considering the softening of cracking concrete	Considered
Zheng et al. [33]	Elastic mechanics	Double-layer thick-walled cylinder	Smeared crack, considering the softening of cracking concrete	Not considered
Pantazopoulou et al. [34]	Elastic mechanics	Double-layer thick-walled cylinder	Cracking concrete, performance is determined by the stress–strain curve	Not considered
Chernin et al. [35]	Elastic mechanics	Double-layer thick-walled cylinder	Circumferential elastic modulus changes with radius	Not considered
Maaddawy et al. [36]	Elastic mechanics	Thick-walled cylinder	Not considered	Not considered
Lu et al. [37]	Elastic mechanics	Thick-walled cylinder	Not considered	Considered
Kim et al. [38]	Elastic mechanics	Thick-walled cylinder	Not considered	Not considered
Malumbela et al. [39]	Elastic mechanics	Thick-walled cylinder	Not considered	Not considered
Li et al. [40]	Muskhelishvili complex function	Infinite plane	Not considered	Not considered

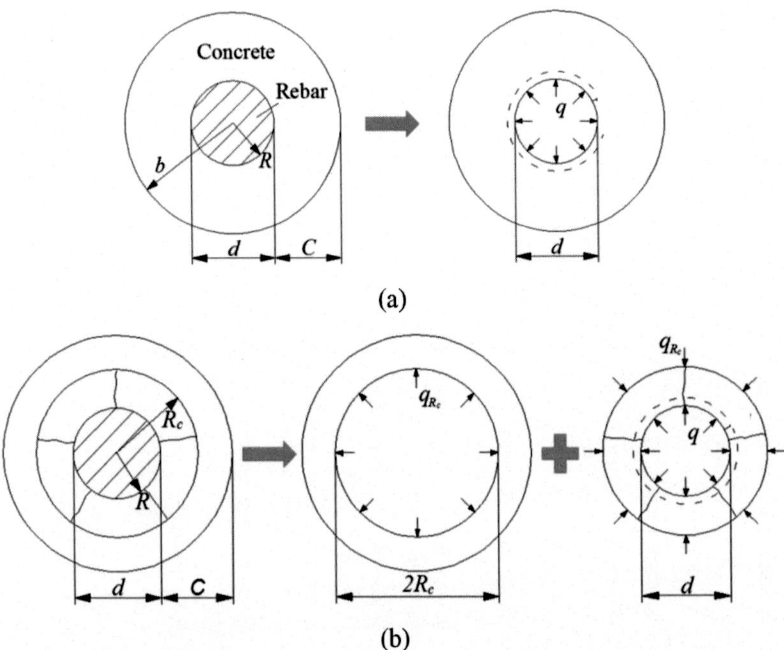

FIGURE 1.5 Corrosion-induced concrete cracking model. (a) Thick-walled cylinder model. (b) Double-layer thick-walled cylinder model.

cracking of the surrounding concrete but also participates in the mechanical interaction during the cracking process. The expansion coefficient and elastic modulus of steel corrosion are important parameters for the corrosion-induced concrete cracking model. These parameters are carefully investigated in chapter "Expansion Coefficients and Modulus of Steel Corrosion" and are used later in the concrete cracking prediction model in chapters "Damage Analysis and Cracking Model of Reinforced Concrete Structures with Rebar Corrosion," "Critical Thickness of Rust Layer at Concrete Surface Cracking," and "Corrosion-Induced Concrete Cracking Model Considering Corrosion Products-Filled Paste."

Table 1.2 shows that the aforementioned theoretical models are all based on the assumption of uniform corrosion. However, this is rarely the case in practice, and the corrosion around the rebar perimeter is not uniform. Therefore, the variability of the thickness of the corrosion layer deposited on the circumference of the steel bar is quantitatively investigated in chapter "Non-Uniform Distribution of Rust Layer Around Steel Bar in Concrete" to improve the mathematical corrosion-induced concrete cracking model.

TABLE 1.3 Rust Filling in Corrosion-Induced Cracks

Reference	Method	Stage 3: Rust Filling in Corrosion-Induced Cracks
Wong et al. [23]	Dry–wet cycle accelerated corrosion	Studied, but no qualitative conclusion
Michel et al. [24]	DC current accelerated corrosion	No rust filling the cracks in the process of electrifying corrosion
Zhao et al. [25]	DC current accelerated corrosion	No rust filling the cracks
Zhao et al. [26]	Dry–wet cycle accelerated corrosion	No rust filling the cracks before concrete surface cracking

1.3.4 Rust Filling in Corrosion-Induced Cracks

The rust filling in corrosion-induced cracks has been studied by several researchers, as shown in Table 1.3. The authors studied the CP in corrosion-induced cracks of reinforced concrete specimens and found that, before concrete surface cracking during the corrosion process, the rust can neither fill the cracks nor penetrate the concrete adjacent to cracks. According to this observation, the rust filling the crack does not need to be considered for the corrosion-induced concrete surface cracking model, that is, δ_{crack} during stage 3 does not need to be included in Eq. (1.1). chapters "Critical Thickness of Rust Layer at Concrete Surface Cracking" and "Rust Distribution in Corrosion-Induced Cracking Concrete" present the detailed experimental process and results.

1.4 CONTENTS OF THIS BOOK

Chapter "Introduction" reviews a number of empirical and analytical models to describe the cracking process of concrete induced by steel corrosion. The predictable performance and the comparison of empirical models are discussed, and achievements in the three-stage analytical cracking model, the corrosion products filling stage, concrete cover stressing stage, and cracking stage, are reviewed.

Chapter "Steel Corrosion in Concrete" introduces the mechanisms of steel corrosion in concrete, which is an electrochemical process whereby the iron is removed from the steel being corroded and dissolved as the ferrous ions in the limited volume of pore solution present in the pores of the concrete surrounding the steel.

Considering that steel corrosion not only directly induces the cracking of the concrete cover but also participates in the mechanical interactions during the concrete cover cracking process, chapter "Expansion Coefficients and Modulus of Steel Corrosion Products" carefully investigates the composition, expansion coefficient, and elastic modulus of steel corrosion.

With these parameters of steel corrosion, in chapter "Damage Analysis and Cracking Model of Reinforced Concrete Structures with Rebar Corrosion" the damage mechanics and elastic mechanics are utilized to study the stresses and strains in the surrounding concrete during the corrosion-induced cracking process and the amount of steel corrosion at the cracking of the concrete cover by taking into account the mechanical properties of both uncracked and cracked concrete, as well as the rust products.

In chapter "Critical Thickness of Rust Layer at Concrete Surface Cracking," the electrochemically corroded reinforced concrete specimens are observed by a digital microscope and a scanning electron microscope (SEM) to study the critical thickness of the corrosion layer at concrete surface cracking and the rust distributed at the steel–concrete interface and in corrosion-induced cracks.

A reinforced concrete specimen subjected to wetting–drying cycles was investigated in chapter "Rust Distribution in Corrosion-Induced Cracking Concrete." The rust distribution was observed by digital microscopy and SEM. This chapter confirms the finding in chapter "Critical Thickness of Rust Layer at Concrete Surface Cracking" that the rust did not fill the corrosion-induced cracks in the concrete cover before concrete surface cracking. Thus, the rust filling the corrosion-induced cracks does not need to be considered in the corrosion-induced concrete surface cracking model.

A Gaussian function is proposed to describe the nonuniform spatial distribution of the corrosion product in chapter "Non-Uniform Distribution of Rust Layer Around Steel Bar in Concrete." This chapter also discusses the physical meaning of the parameters in a Gaussian model and investigates the relationships among these parameters by measuring the thickness of the nonuniform corrosion layer around the perimeter of the steel bar in reinforced concrete specimens.

The shape of the corrosion-induced cracks in the concrete cover is observed and investigated in chapter "Crack Shape in Corrosion-Induced Cracking Concrete Cover," and a linear model is proposed to describe the variation of the total circumferential crack width along the radial direction in the concrete cover. The parameters in the model can be interpreted physically to describe various characteristics of the cracks.

Chapter "Development of the Corrosion Products-Filled Paste at the Steel/Concrete Interface" carefully investigates the rust distribution at the steel–concrete interface and finds the penetration of corrosion products into the porous zone of concrete and the formation of a corrosion layer at the steel–concrete interface process simultaneously after the initiation of steel corrosion, although not separately, as previously assumed.

Finally, in chapter "Steel Corrosion-Induced Concrete Cracking Model," an improved corrosion-induced cracking model is proposed, and it considers the phenomena, found in chapters "Rust Distribution in Corrosion-Induced Cracking Concrete" and "Development of the Corrosion Products-Filled Paste at the Steel/Concrete Interface," that the corrosion layer accumulation and corrosion products filling pores occur simultaneously in concrete. The time from corrosion initiation to concrete surface cracking is discussed. Further improvements to the corrosion-induced cracking model are also discussed in this chapter.

REFERENCES

[1] Bamforth PB. Enhancing reinforced concrete durability guidance on selecting measures for minimising the risk of corrosion of reinforcement in concrete. Concrete Soc Tech Rep 2004;61.

[2] Tuutti K. Corrosion of steel in concrete, Swedish Cement and Concrete Research Institute, Fo 4.82, Stockholm (1982).

[3] Alonso C, Andrade C, Rodriguez J, et al. Factors controlling cracking of concrete affected by reinforcement corrosion. Mater Struct 1998;31(7):435–41.

[4] Lindvall A. DuraCrete—probabilistic performance based durability design of concrete structures, Brite EuRam III, Project BE95-1347, Document BE95-1347/R17 (2000).

[5] Andrade C, Alonso C, Molina FJ. Cover cracking as a function of bar corrosion: part I-Experimental test. Mater Struct 1993;26(8):453–64.

[6] Rodriguez J, Ortega LM, Casal J, et al. Corrosion of reinforcement and service life of concrete structures. Durability Build Mater Components 1996;7(1):117–26.

[7] DuraCrete. Modeling of degradation. brite-euram-project BE95-1347/R4-5; 1998.

[8] Zhang WP. The methods of damage prediction and durability estimation for corrosion of reinforcement in concrete structures[D]. Master's degree thesis of Tongji University, Shanghai, P.R. China, 1999 [in Chinese].

[9] Oh BH, Kim KH, Jang BS. Critical corrosion amount to cause cracking of reinforced concrete structures. ACI Mater J 2009;106(4):333–9.

[10] Webster MP, Clark LA. The structural effect of corrosion–an overview of the mechanism[J]. Proceedings of the Concrete Communication, Birmingham, UK; 2000. p. 409–421.

[11] Wang HL, Jin WL, Sun XY. Corrosion-induced cracking model of reinforced concrete cover based on fracture mechanics. Shuili Xuebao 2008;39(7):863–9 [in Chinese].

[12] Vu K, Stewart MG, Mullard J. Corrosion-induced cracking: experimental data and predictive models. ACI Struct J 2005;102(5):719–26.

[13] Mullard JA, Stewart M.G. Corrosion-induced cover cracking of RC structures: new experimental data and predictive modelsl NOVA. The University of Newcastle's Digital Repository 2009.

[14] Vidal T, Castel A, Francois R. Analyzing crack width to predict corrosion in reinforced concrete. Cement Concrete Res 2004;34(1):165–74.

[15] Zhang R, Castel A, Franois R. Concrete cover cracking with reinforcement corrosion of RC beam during chloride-induced corrosion process. Cement Concrete Res 2010;40 (3):415–25.

[16] Liu Y, Weyers RE. Modeling the time-to-corrosion cracking in chloride contaminated reinforced concrete structures. ACI Mater J 1998;95(6).

[17] Amin J, Ueli A, Bryan A, Bernhard E. Modeling of corrosion-induced concrete cover cracking: a critical analysis. Construction Build Mater 2013;42:225–37.

[18] Asami K, Kikuchi M. In-depth distribution of rusts on a plain carbon steel and weathering steels exposed to coastal-industrial atmosphere for 17 years. Corrosion Sci 2003;45 (11):2671–88.

[19] Duff S, Morris W, Raspini I, et al. A study of steel rebars embedded in concrete during 65 years. Corrosion Sci 2004;46(9):2143–57.

[20] Chitty W, Dillmann P, L'Hostis V, et al. Long-term corrosion resistance of metallic reinforcements in concrete—a study of corrosion mechanisms based on archaeological artefacts. Corrosion Sci 2005;47(6):1555–81.

[21] Car S, Nguyen QT, L'Hostis V, et al. Mechanical properties of the rust layer induced by impressed current method in reinforced mortar. Cement Concrete Res 2008;38(8):1079–91.

[22] Jaffer SJ, Hansson CM. Chloride-induced corrosion products of steel in cracked-concrete subjected to different loading conditions. Cement Concrete Res 2009;39(2):116–25.

[23] Wong HS, Zhao YX, Karimi AR, et al. On the penetration of corrosion products from reinforcing steel into concrete due to chloride-induced corrosion. Corrosion Sci 2010;52 (7):2469–80.

[24] Michel A, Pease BJ, Geiker MR, et al. Monitoring reinforcement corrosion and corrosion-induced cracking using non-destructive x-ray attenuation measurements. Cement Concrete Res 2011;41(11):1085–94.

[25] Zhao YX, Yu J, Wu YY, et al. Critical thickness of rust layer at inner and out surface cracking of concrete cover in reinforced concrete structures. Corrosion Sci 2012;59:316–23.

[26] Zhao YX, Wu YY, Jin WL. Distribution of millscale on corroded steel bars and penetration of steel corrosion products in concrete. Corrosion Sci 2013;66:160–8.

[27] Zhao YX, Ding HJ, Jin WL. Development of the corrosion-filled paste and corrosion layer at the steel/concrete interface. Corrosion Sci 2014;87:199–210.

[28] Bazant ZP. Physical model for steel corrosion in concrete sea structures—application. J Struct Div 1979;105 [ASCE 14652 Proceeding].

[29] Bhargava K, Ghosh AK, Mori Y, et al. Analytical model for time to cover cracking in RC structures due to rebar corrosion. Nucl Eng Design 2006;236(11):1123–39.

[30] Zhao YX, Jin WL. Modeling the amount of steel corrosion at the cracking of concrete cover. Adv Struct Eng 2006;9(5):687–96.

[31] Li C, Melchers RE, Zheng J. Analytical model for corrosion-induced crack width in reinforced concrete structures. ACI Struct J 2006;103(4):479–87.

[32] Yu J. Damage analysis and experimental study of reinforced concrete structures with rebar corrosion[D]. Hangzhou, Zhejiang University, 2013 [in Chinese].

[33] Zheng JJ, Zhou XZ, Li CQ. Analytical solution to corrosion damage of reinforced concrete structures. Shuili Xuebao 2004;(12):62–8 [in Chinese].

[34] Pantazopoulou SJ, Papoulia KD. Modeling cover-cracking due to reinforcement corrosion in RC structures. J Eng Mech 2001;127(4):342–51.

[35] Chernin L, Val DV, Volokh KY. Analytical modelling of concrete cover cracking caused by corrosion of reinforcement. Mater Struct 2010;43(4):543–56.

[36] El MT, Soudki K. A model for prediction of time from corrosion initiation to corrosion cracking. Cement Concrete Composites 2007;29(3):168–75.

[37] Lu C, Jin W, Liu R. Reinforcement corrosion-induced cover cracking and its time prediction for reinforced concrete structure. Corrosion Sci 2011;53(4):1337–47.

[38] Kim KH, Jang SY, Jang BS, et al. Modeling mechanical behavior of reinforced concrete due to corrosion of steel bar. ACI Mater J 2010;107(2):106–13.

[39] Malumbela G, Alexander M, Moyo P. Model for cover cracking of RC beams due to partial surface steel corrosion. Construction Build Mater 2011;25(2):987–91.

[40] Li S, Wang M, Li S. Model for cover cracking due to corrosion expansion and uniform stresses at infinity. Appl Math Model 2008;32(7):1436–44.

Chapter 2

Steel Corrosion in Concrete

Chapter Outline

2.1 INTRODUCTION

Steel corrosion in any environment is an electrochemical process in which iron (Fe) is removed from the steel being corroded and is dissolved into the surrounding solution; it then appears as ferrous ions (Fe^+). For steel embedded in concrete, the dissolution takes place in the limited volume of water solution present in the pores of the concrete surrounding the steel.

The ferrous ions dissolved in the concrete pore solutions usually react with hydroxide ions (OH^-) and dissolved oxygen molecules (O_2) to form one or a combination of several varieties of rust, which is a solid by-product of the corrosion reaction. The rust is usually deposited on the interface of the concrete and steel bars. Its formation within this restricted space initiates expansive stresses that crack the concrete cover. These cracks align with reinforcing bars and occur over them; this is related to a serious durability concern.

The steel corrosion not only causes this type of concrete cover cracking but also participates in the mechanical interactions that are induced by the rust volume expansion between the concrete cover and the steel. Therefore, it is a very important factor in the corrosion-induced concrete cracking process. This chapter briefly introduces the mechanisms of steel corrosion in concrete.

Steel Corrosion-Induced Concrete Cracking. DOI: http://dx.doi.org/10.1016/B978-0-12-809197-5.00002-5

2.2 MECHANISMS OF STEEL CORROSION IN CONCRETE

2.2.1 Corrosion Process

The corrosion process is illustrated in Fig. 2.1. This shows the formation of anodic and catholic sites on a steel bar. When steel corrodes in concrete, it dissolves in the pore solutions and gives up elections, and steel corrosion (Fe^+) occurs at the anode [1]:

$$Fe \rightarrow Fe^{2+} + 2e^- \tag{2.1}$$

Electrons are freed and flow though the steel to the cathode [1]:

$$2e^- + H_2O + O_2/2 \rightarrow 2OH^- \tag{2.2}$$

The electrons are consumed in the reduction processes leading to the formation of hydroxide ions. The permeability of the concrete cover allows oxygen and moisture to penetrate the cover to feed the cathodic reaction. Sufficiently saturated moist concrete serves as an electrolyte.

The anodic and cathodic reactions shown in Eqs. (2.1) and (2.2) are only the first steps in the process of creating corrosion products. Several more stages occur for "rust" to form, and its occurrence can be expressed in several ways. One typical way is when the ferrous hydroxide becomes the ferric hydroxide and then becomes the hydrated ferric oxide or rust, as presented here [1]:

$$Fe + 2OH^- \rightarrow Fe(OH)_2 \quad \text{Ferrous hydroxide}$$

$$4Fe(OH)_2 + O_2 + 2H_2O \rightarrow 4Fe(OH)_3 \quad \text{Ferric hydroxide}$$

$$Fe(OH)_3 \rightarrow Fe_2O_3 \bullet H_2O + H_2O \quad \text{Hydrated ferric oxide rust}$$

Underrated ferric oxide Fe_2O_3 has a volume of approximately twice that of the steel it replaces when fully dense. When it becomes hydrated, it swells even more. This means that the volume increase at the steel–concrete interface is two-times or more greater than the original volume, leading to the cracking and spalling observed as the consequence of corrosion of steel in concrete, the red/brown, brittle, flack rust on the bar, and the rust stains observed at cracks in the concrete. The different types of corrosion products are introduced in Section 2.4.

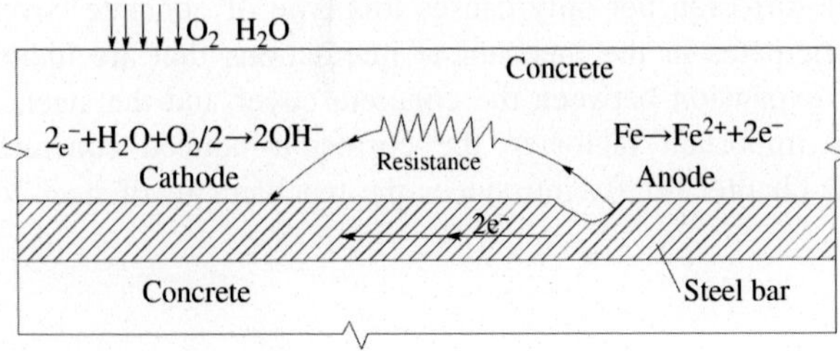

FIGURE 2.1 The anodic and cathodic reactions.

2.2.2 Corrosion Rate

The anodic and cathodic processes lead to an accumulation of positive and negative charges, respectively, but this is not sustained. The hydroxide ions diffuse toward the anode, where they meet the ferrous ions, and the resulting combination causes electrical neutralization if the anodic and cathodic processes are coupled together in the form of a corrosion cell with no excess electrons. If there is no external source of electrons, then the electrons produced by oxidation will be fully consumed by reduction. Thus, the oxidation rate at the anode and the reduction rate at the cathode must be equal, and this equality controls the corrosion rate. Therefore, the rate of electron flow reflects the rate of corrosion.

One important factor controlling the rate of corrosion is the availability of dissolved oxygen surrounding the cathodic areas [2] (Eq. 2.2). Oxygen is consumed in the cathodic reaction. If its supply in the solution surrounding the cathodic areas of the metal is not continuously supplied, then the corrosion reactions may be restricted. One way this can occur is when the surface of the steel is surrounded by a concrete cover, which slows the diffusion of oxygen from the surrounding environment, as illustrated in Fig. 2.1. In this case, the rate of corrosion becomes "diffusion controlled," which means it is regulated by the rate of oxygen diffusion through the concrete cover. The regulation of the corrosion rate controlled by slow diffusion of oxygen produces a significant reduction in the potential difference between the anodic and cathodic areas. Such an effect is called a "polarizing effect," and the process is called "polarization."

The other important factor in the steel corrosion process in concrete is the limitation of the ionic current flow thought the pores of the concrete surrounding the steel [2]. If the flow rate of the charge-carrying ionic species is slow, then the corrosion reactions can proceed only at a slow rate. This happens when the electrical resistance of the concrete surrounding the steel between the anode and the cathode is high. In practice, the phenomenon of steel corrosion does not easily occur in dry concrete due to the high electrical resistance of the dry concrete cover. Therefore, the measurement of the electrical resistance of concrete cover can sometimes serve as an indication of how fast the corrosion reactions proceed.

Another type of process that limits the rate of corrosion is passivation.

2.2.3 Passivation

Appreciable steel corrosion does not usually happen in dense concrete due to the alkalinity of the concrete; the pore liquid phase of hardened concrete can have a pH value as high as 13 due to the presence of dissolved NaOH and KOH [3]. When steel is exposed to an alkaline condition at a pH value more than 11.5 in the presence of dissolved oxygen, steel can react with oxygen to form thin layers of insoluble metal oxide/hydroxide on its surface. This passive layer is a thin (approximately 10 nm), dense, impenetrable film that serves as a barrier to the anodic iron dissolution process and reduces the corrosion rate of the steel to an imperceptible level, corresponding to an average rate of reduction in metal thickness of less than 1 μm/year [3]. If this passive film remains effective, then the corrosion is so slow that, for practical purposes, the steel can be considered noncorroding. This passivate film is the engineer's dream coating because it forms itself and maintains and repairs itself as long as the passivating (alkaline) environment is there to regenerate it if the film is damaged.

The alkaline condition of concrete creates a stable passive layer on the surface of embedded steel bars unless the environmental condition changes. In reality, this alkaline environment of concrete is not always maintained. Two processes can break down the passivating environment in concrete: carbonation and chloride attack.

2.3 STEEL CORROSION INDUCED BY CARBONATION OR CHLORIDE ATTACK

The main causes of steel corrosion in concrete are carbonation and chloride attack. These two mechanisms are unusual in that they do not attack the integrity of the concrete. Instead, aggressive chemical species pass though the pores in the concrete and attack the steel. This is unlike normal deterioration processes due to chemical attack on concrete. Other acids and aggressive ions such as sulfate destroy the concrete before the steel is affected. Most forms of chemical attack are concrete problems before corrosion problems. Carbon dioxide and chloride ions penetrate the concrete without significantly damaging it; they attack the steel but not the concrete.

The service life of reinforced concrete structures can be divided into two distinct phases, as illustrated by Tuutti's model [4] in Fig. 2.2.

The first phase is the initiation of corrosion, whereby the reinforcement is passive but phenomena that can lead to loss of passivity occur, such as carbonation or chloride perpetration in the concrete cover. The duration of the initiation phase depends on the cover depth and the penetration rate of the aggressive agents. The influence of the concrete cover is obvious, and design codes normally define minimum cover depths. The rate of ingress of the aggressive agents depends on the porosity of the concrete and the microclimatic conditions at the concrete surface.

The second phase is propagation of corrosion that begins when the steel is depassivated and eventually results in a limit state being reached, which is usually identified by failure of serviceability associated with cracking or spalling of the concrete cover. This book focuses on the propagation period related to corrosion-induced concrete cracking.

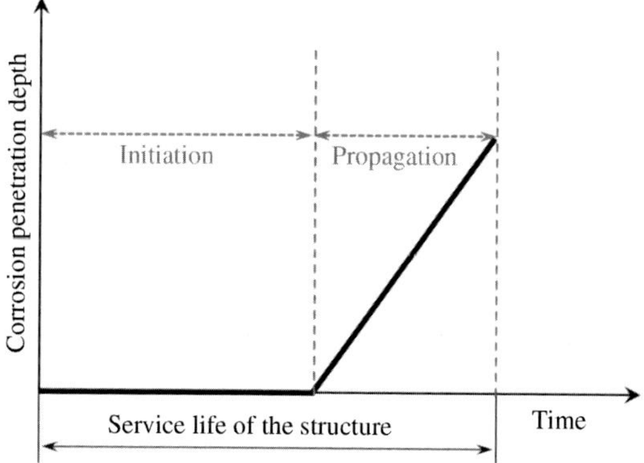

FIGURE 2.2 Initiation and propagation periods for steel corrosion in concrete [4].

2.3.1 Carbonation-Induced Corrosion

Carbonation begins at the surface of concrete and moves gradually toward the inner zones. The alkalinity of concrete is neutralized by CO_2 from the atmosphere and the pH of the pore liquid of the concrete gradually decreases to near-neutral levels, as described by Eqs. (2.3) and (2.4). In this circumstance, the passive film is no longer stable.

$$CO_2 + H_2O \rightarrow H_2CO_3 \tag{2.3}$$

$$Ca(OH)_2 + H_2CO_3 \rightarrow CaCO_3 + 2H_2O \tag{2.4}$$

After the full-depth of the concrete cover is carbonated, the alkaline environment protecting the passive film no longer exists and corrosion initiates on the surface of the steel bar. In this circumstance, the steel surface may be viewed as consisting of clusters of local microcells (ie, a combination of adjacent potentially anodic and cathodic sites) [5]. When the environment is moist enough, the currents will flow and the local cells will become active, creating the onset of corrosion. In this process, this corrosion generally takes place more or less uniformly over the whole steel surface. Thus, this corrosion is often termed "general corrosion," which is discussed in this book.

2.3.2 Chloride-Induced Corrosion

Alkalinity in the concrete pores is neutralized by carbonation, but the depassivation mechanism for chloride attack is somewhat different. The chloride ions attack the passive film, but no overall decrease in pH occurs, unlike carbonation. Chlorides act as catalysts to corrosion when there is sufficient concentration at the steel bar surface to break down the passive film. The chloride ions are not consumed in the process but rather help to break down the passive film on the steel and allow the corrosion process to proceed quickly. The equation can be briefly expressed as follows:

$$Fe^{2+} + 2Cl^- + 4H_2O \rightarrow FeCl_2 \cdot 4H_2O$$

$$FeCl_2 \cdot 4H_2O \rightarrow Fe(OH)_2 + 2Cl^- + 2H^+ + 2H_2O$$

As mentioned previously, effective passivation can be achieved and maintained for conventional steel embedded in concrete. The concentration of the hydroxide ions in the pore solution of the concrete in contact with the steel is usually sufficiently high to maintain this passivation film. A few chloride ions in the pore solution will not break down the passive film. However, if there is a sufficient concentration, then the chloride ions can depassivate steel and cause pitting corrosion to occur. The determination of the chloride threshold values under various circumstances is a very complicated problem. The complexity and variety of the "chloride threshold" are due to several reasons: concrete pH varies with the type of cement and the concrete mixture; chlorides can be chemically and physically bound in concrete; in very dry concrete or surface-treated concrete, corrosion may not occur even at very high chloride ion concentrations because not enough moisture or oxygen is present in concrete to activate the corrosion reaction; and corrosion can be suppressed if there is total water saturation due to oxygen starvation. Therefore, corrosion can be observed as a threshold level of 0.2% chloride by weight of cement if the water and oxygen are available or up to 1.0% or more if water and oxygen are excluded [1].

After the accumulation of the chloride ions at the surface of the steel bar exceeds the chloride threshold, the pitting corrosion, as schematically illustrated in Fig. 2.3, occurs at some sites on the steel surface (normally thought to be a void in the cement paste or a sulfide inclusion in the steel [1])

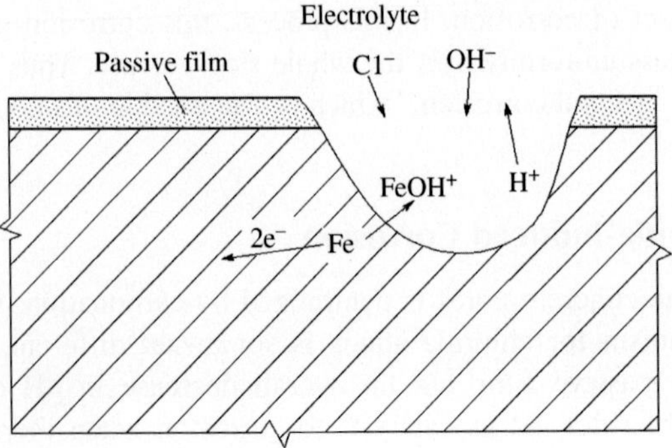

FIGURE 2.3 Pitting attack in a steel bar.

where the passive film is more vulnerable to attack and an electrochemical potential difference attracts chloride ions. After corrosion starts, acids are present; a pit forms and the rust generates over the pit, concentrating the acid. Under this circumstance, the anodes and cathodes are often well separated with large cathodic areas supporting small, concentrated anodic areas. This is known as the macro cell phenomenon.

The loss of the cross-sectional area in pitting corrosion occurs rapidly and critically reduces the load-bearing capacity of the reinforced concrete members. Unlike carbonation-induced corrosion, in which cracking and spalling can serve as an early indicator of a problem, chloride-induced pitting corrosion may be quite advanced before evidence is apparent [5].

Moreover, combined with the hygroscopicity of the chloride, the ohmic resistance of the concrete cover could be reduced and the galvanic corrosion activity could be accelerated.

2.4 CORROSION PRODUCTS

Steel corrosion products (ie, iron oxides) are common compounds that are widespread in nature. There are 16 iron oxides, as listed in Table 2.1. They are composed of Fe together with O and/or OH. The individual iron oxides [6] are described briefly.

Goethite, α-FeOOH, is one of the most thermodynamically stable iron oxides at ambient temperatures; therefore, it is the first oxide to form and also the end member of many transformations. In massive crystal aggregates, goethite is dark brown or black, but the powder is yellow and is responsible for the color of many rocks, soils, and ocher deposits. Industrially, goethite is an important pigment.

TABLE 2.1 Different Types of Iron Oxides [6]

Oxide Hydroxides and Hydroxides	Oxides
Goethite α-FeOOH	Hematite α-Fe_2O_3
Lepidocrocite γ-FeOOH	Magnetite Fe_3O_4
Akaganeite β-FeOOH	Maghemite γ-Fe_2O_3
Schwertmannite	
δ-FeOOH	
Feroxyhyte (δ'-FeOOH)	
Ferrihydrite	
Bernalite $Fe(OH)_3$	
$Fe(OH)_2$	

The orange *lepidocrocite*, γ-FeOOH, is typically formed in nature (ie, in soils, biota, and rust) as an oxidation product of Fe^{2+}.

Akaganeite, β-FeOOH, occurs rarely in nature. It is found mainly in Cl-rich environments such as offshore structures. Its color is brown to bright yellow. In a related compound, *schwertmannite*, the chloride ions have been replaced by sulfate ions. It is also found in nature (eg, in acid mine waters) as an oxidation product of pyrite.

δ-FeOOH (synthetic) and its poorly crystalline mineral form, *feroxyhyte* (δ'-FeOOH), are ferromagnetic materials. These compounds are reddish brown.

Ferrihydrite, a reddish-brown mineral, is widespread in surface environments. Unlike the other iron oxides, it is poorly ordered and, unless stabilized in some way, transforms into more stable members of the group.

Bernalite, $Fe(OH)_3$, is a greenish iron oxide that, to date, has only been found as a mineral sample in a museum.

$Fe(OH)_2$ does not exist as a mineral. Pure $Fe(OH)_2$ is white. However, it is readily oxidized; on oxidation, it develops into greenish-blue so-called green rust or, on further oxidation, into black magnetite.

Hematite, α-Fe_2O_3, is the oldest known Fe oxide mineral. Its color is red if finely divided or black or sparkling gray if coarsely crystalline. Like goethite, hematite is extremely stable and is often the end member of transformations of other iron oxides.

Magnetite, Fe_3O_4, is a black ferromagnetic mineral containing both Fe^{II} and Fe^{III}. Magnetite is an important ore.

Maghemite, γ-Fe_2O_3, is a red-brown ferromagnetic material. It occurs in soils as a weathering product of magnetite and as the product of heating other Fe oxides in the presence of organic matter. Maghemite is an important magnetic pigment.

β-Fe_2O_3 and ε-Fe_2O_3 are rare compounds that have only been synthesized in the laboratory.

Wüstite, FeO, is a black iron oxide containing only divalent Fe. Wüstite is an important intermediate in the reduction of iron ores.

Different types of iron oxides transform under different environments, as briefly illustrated in Fig. 2.4. As stated previously, Fe_2O_3 and α—FeO(OH) are the most stable iron oxides and are often the end members of the transformations [7,8].

The selected properties of the iron oxides, which are related to the corrosion-induced cracking process introduced in this book, are listed in Table 2.2. The expansion coefficient is the ratio of the volumes between the iron oxides and its original steel.

The composition of steel corrosion that may occur in concrete structures is discussed in detail in chapter "The Expansion Coefficients and Modulus of Steel Corrosion."

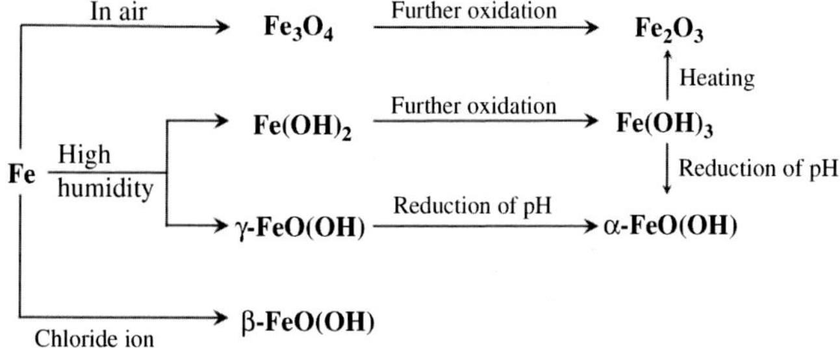

FIGURE 2.4 Transformation of iron oxides.

TABLE 2.2 Selected Properties of the Iron Oxides [6]

Corrosion Products	Chemical Formula	Density ($g \cdot cm^3$)	Color
Goethite	α-FeOOH	4.26	Yellow-brown
Lepidocrocite	γ-FeOOH	4.09	Orange
Akaganeite	β-FeOOH	3.56	Yellow-brown
Feroxyhyte	δ′-FeOOH	4.20	Red-brown
Ferrihydrite		3.96	Red-brown
Hematite	α-Fe_2O_3	5.26	Red
Magnetite	Fe_3O_4	5.18	Black
Maghemite	γ-Fe_2O_3	4.87	Reddish-brown
Wüstite	FeO	5.9–5.99	Black

2.5 STEEL CORROSION-INDUCED CONCRETE DAMAGE

In most industries, steel corrosion is a concern because of the wastage of metal leading to structural damage, such as collapse of steel structures and perforation of containers. However, most problems with corrosion of steel in concrete are not due to loss of steel but rather to the growth of the oxide, which leads to cracking and spalling of the concrete cover.

Structural collapses of reinforced concrete structures due to steel corrosion are rare; the observation of cracking, rust staining, and spalling of the concrete cover usually appears well before a reinforced concrete structure is at risk [1,9]. Field studies also suggest that cracking and spalling are of most concern to asset owners [10]. This is broadly consistent with the general observation that the service life of a structure is reduced considerably if crack widths exceeding 0.3–0.5 mm are not repaired. If not repaired, then

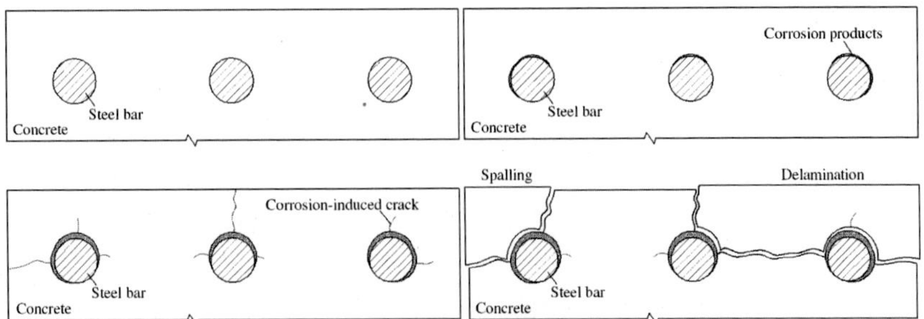

FIGURE 2.5 Stages in corrosion-induced damage. (a) Passive rebar. (b) Corrosion initiation and growth. (c) Further corrosion and cracking propagation. (d) Spalling/delamination.

corrosion will eventually lead to structural distress over time. Therefore, corrosion-induced cracking and delamination are increasingly considered influential modes of failure for the estimation of life cycle costs and optimization of repair and maintenance strategies for RC structures.

The most common problems caused by steel corrosion are cracking and spalling of concrete cover, as illustrated in Fig. 2.5. The important factors of steel corrosion in concrete compared to most other corrosion problems are the volume of oxide and where it is formed. A dense oxide usually has two-times or more of the volume of its basic steel. For steel in concrete, there are two problems. The main problem is that the pore solution is static and cannot move the oxide away from the steel surface, meaning that the corrosion products are deposited at the steel–concrete interface. The second problem is that the corrosion products occupies a large volume, normally two- to four-times that of the original steel. As a result, coupled with the low tensile strength of concrete, the corrosion products break up the corrosion. It has been suggested [1] that less than 100 μm of steel radial loss starts cracking the concrete cover. The actual amount needed to crack concrete depends on the geometry in terms of cover, proximity to corners, rebar diameter, and spacing. Generally, if the concrete cover thickness is relatively small with respect to the rebar spacing, it is likely that planar cracks with 45-degree angles will develop. When these cracks reach the surface of the concrete, spalling occurs. If the reinforcing bars are closely spaced, then the cracks tend to develop and progress across the plane of the reinforcing bars, causing delamination. There is further discussion of corrosion-induced damage in the following chapters.

2.6 CONCLUSIONS

1. The electrochemical mechanisms of steel corrosion in concrete are briefly introduced in this chapter. The factors controlling the rate of corrosion are the availability of dissolved oxygen and the electrical resistance of concrete surrounding the steel.

2. The alkaline condition of concrete leads to a stable passive layer, which is a thin, dense, impenetrable film. This passive film serves as a barrier to corrosion and reduces the corrosion rate of the steel to an imperceptible level.
3. The corrosion initiation and propagation processes under carbonation and chloride ingress environments are discussed. Carbonation-induced corrosion is often "general corrosion," whereas chloride attack causes "pitting corrosion."
4. Different types of iron oxides and their transformations under different conditions are briefly introduced. Fe_2O_3 and α—FeO(OH) are the most stable iron oxides and are often the end members of the transformations. The selected properties of the iron oxides, which are related to the introduced corrosion-induced cracking process, are listed.

REFERENCES

[1] Broomfield JP. Corrosion of steel in concrete: understanding, investigation and repair. London and New York, NY: E & FN Spon Press; 1997.

[2] Bentur A, Diamond S, Berke NS. Steel corrosion in concrete: fundamentals and civil engineering practice. London and New York, NY: E & FN Spon Press; 1997.

[3] Page CL, Page MM. Durability of concrete and cement composites. Boca Raton, FL: CRC Press; 2007.

[4] Tuutti K. Corrosion of steel in concrete[R]. Report 4, Swedish Cement and Concrete Research Institute, Stockholm, Sweden, April 1982. p. 82

[5] Bertolini L, Elsener B, Pedeferri P, Polder R. Corrosion of steel in concrete: prevention, diagnosis, repair. Weinheim and Cambridge: WILEY-VCH Press; 2003.

[6] Richardson MG. Fundamentals of durable reinforced concrete. London and New York, NY: Spon Press; 2001.

[7] Cornell RM, Schwertmann U. The iron oxides: structure, properties, reactions, occurrence and uses. Weinheim and New York, NY: VCH Press; 1996.

[8] Ishikawa T, Takeuchi K, Kandori K, et al. Transformation of γ-FeOOH to α-FeOOH in acidic solutions containing metal ions. Colloids Surf A Physicochem Eng Asp 2005;266 (1):155–9.

[9] Vu K, Stewart MG, Mullard J. Corrosion-induced cracking: experimental data and predictive models. ACI Struct J 2005;102(5):719–27.

[10] Hartt WL, Lee SK, Costa E. Condition assessment and deterioration rate for chloride contaminated reinforced concrete structures. In: Proceedings of the international seminar on repair and rehabilitation of RC structures: the state of the art, ASCE, 1998, p. 82–104.

Chapter 3

The Expansion Coefficients and Modulus of Steel Corrosion Products

Chapter Outline

3.1 INTRODUCTION

During the concrete cover cracking process, steel corrosion products directly induces the cracking of the concrete cover. It also participates in mechanical interactions that are induced by the volume expansion between the concrete cover and the steel. Therefore, the rust volume expansion coefficient and the mechanical behavior of the rust should be considered in the concrete cracking model.

The expansion coefficient for the corrosion product varies from 2.2 to 6.4 [1]. In previous research of the model of concrete cracking due to corrosion, the values of the rust expansion coefficient varied between 2 and 4 [2–10]. The value of 2.0 was used by Molina [2], Noghabai [3], Lundgren [4], and Wang [5]. The values of 2.7 and 3.0 were used by Lu [6] and Zhao [7] in the analyses of concrete cracking due to corrosion. The values of 2.94 and 3.39 were used by Bhargava [8] and Val [9] in their thermal simulation. The value of 4.0 was used by Liu and Weyers [10] in their concrete cracking model. It has to be pointed out here that all of the values mentioned were assumed by the authors and did not consider the compositions of the corrosion products and the influence of the environments, which in turn may affect the compositions. Caré and Nguyen [11] tried to determine the expansion coefficient for

Steel Corrosion-Induced Concrete Cracking. DOI: http://dx.doi.org/10.1016/B978-0-12-809197-5.00003-7

each corrosion product with a modeling approach considering the varying thickness of a three-phase material (steel/rust/mortar). They obtained results that are similar to the values published in the literature [1]. However, their results did not consider the complex composition of rust. Syed [12] used X-ray diffraction (XRD) to characterize the corrosion products formed on hot-rolled carbon steel and cold-rolled carbon steel after exposure to the open atmosphere for 3 years at 20 stations in the Kingdom of Saudi Arabia and tried to describe the relation between the environment and the composition of rust. That research showed that the different service environments caused various compositions of rust. However, he did not study the expansion coefficients of the corrosion products produced in different environments.

Meanwhile, studies of the modulus of steel corrosion products that can be used in the concrete cracking model remain insufficient. Although Samsonov [13] found that the modulus of iron oxide crystals ranged from 215 to 350 GPa, these values reflected the properties of the rust components (Fe_2O_3 or Fe_3O_4) but not those of rust, which is also influenced by other factors, such as rust composition varieties and porosity, in addition to the properties of the components. Some researchers [2,8,11,14–17] have suggested that the elastic modulus of steel corrosion products should be used in the corrosion-induced concrete cracking model. Molina et al. [2] assumed that rust was elastic and stated that "the properties of the rust can be replaced by a fluid with Poisson's ratio of $\upsilon_r = 0.5$ and bulk modulus of $K_r = 2.0$ GPa." They calculated the elastic modulus using the equation $E_{rust} = 3(1 - 2\upsilon_r)K_r$, where υ_r could be taken as a value slightly lower than 0.5. To investigate the model for time to cover cracking, Bhargava et al. [8] assumed that the mechanical properties of the rust products (ie, the elastic modulus and Poisson ratio), were identical to those of the reinforcing steel (ie, $E_{rust} = 200$ GPa and $\upsilon_r = 0.3$). However, this assumption is implausible because steel is much harder and denser than rust. Based on the theory of elastic mechanics, Caré et al. [11] considered a hollow cylinder composed of a solid phase that was subjected to inner and outer pressures. In this work, the calibrated Young modulus of the rust that was formed in the accelerated corrosion tests was constant, with a value of $E_r = 0.13$ GPa when Poisson ratio was 0.3. Suda et al. [14] estimated that the modulus of elasticity of the corrosion products ranged from 0.1 to 0.5 GPa based on measuring expansive strains in steel plates that were subjected to galvanostatic corrosion and confined by concrete bars. In a later work, Petre-Lazar and Gerard [15] used the XRD to observe the corrosion products formed in an accelerated test. They concluded that the rust was a cohesionless assemblage of incompressible crystals. This conclusion was based on a qualitative description of the properties of the rust because there were no values provided. Based on the finding of Petre-Lazar and Gerard, Lundgren [16] developed the strain formula of the corrosion layer regarding the two-dimensional geometric relationship of three areas: the rebar, the rust, and the concrete. Stress was evaluated by a combination of experimental results and analyses. The stress–strain curve was obtained from the results using the equation $\sigma_r = K_r \cdot \varepsilon_r^p$, where $K_r = 14$ GPa and $p = 7.0$. However, little data described the low-stress region in Lundgren's work, whereas some work [10,18–21] indicated that the pressure due to corrosion at the concrete and steel interface might be only several MPa. Ouglova et al. [17] prepared particles of rust with sizes varying from 0.4 to 8.7 μm (mostly lepidocrocite). The Young modulus of the particles was found to be approximately 360 GPa, which was higher than that of steel ($E_s = 200$ GPa). The modulus of the rust material shown in the aforementioned literature varied from 100 MPa to 360 GPa.

Therefore, this chapter studies the values of the expansion coefficient and modulus of rust, which could be used in the corrosion-induced concrete cracking model.

3.2 EXPANSION COEFFICIENT OF STEEL CORROSION PRODUCTS

3.2.1 Experimental Program

Eight types of rust used in this study are listed in Table 3.1. They were characterized by using a Rigaku D/max-ra XRD diffractometer with a Cu X-ray target under the condition of 40 kV and 100 mA. Data from the experiment were collected over a 2θ range of 10–80° with a wavelength of 1.5406. The window was fixed and all rust samples turned with a speed of 4°/min.

All rust samples were also investigated with a thermogravimetric (TG) device (WRT-3P) and differential thermal analysis (DTA) device (CRY-2P). The former device was used to describe the relation between the temperature and the sample quality, whereas the latter one recorded the relation between the temperature and the sample thermal energy [22]. The samples were heated in air at a heating rate of 10°C/min. TG and DTA curves (using α-Al2O3 as a reference) were recorded simultaneously from room temperature to 1000°C.

3.2.2 Tested Results

3.2.2.1 X-Ray Diffraction

Fig. 3.1 illustrates the XRD patterns for eight rust samples. A peak marked with a variety of compositions means that at least one of these products existed at this position.

The compositions of Fe_2O_3, Fe_3O_4, α-FeOOH, β-FeOOH, and γ-FeOOH were identified qualitatively in all rust samples as shown in Fig. 3.1. Although there were peaks of akaganeite in all XRD spectra, β-FeOOH was overlapped by other compounds and β-FeOOH was still considered to exist in all the rust samples. This is because β-FeOOH existed in the rust generated from the environments containing chlorine ions [23–25], and all

TABLE 3.1 Details of the Rust Samples

Samples	A	B	C	D
Source	An open-air ironwork in Ningbo (a coastal city in eastern China)	A facility[a] in a factory for the treatment of urban sewage in Shanghai, China	An ironwork[b] at a waste yard in Gansu province (western China)	An iron rack in the durability laboratory of Zhejiang University, Hangzhou, China
Collecting method	Knocking and peeling	Peeling	Peeling	Peeling
Conservation method	Packed in bags	Packed in bags	Packed in bags	Packed in bags
Corroding time	8 years	5 years	10 years	5 years
Conserving time	6 months	5 months	2 months	1.5 months
Classical environment[c]	Moderate humidity (XD1)	Cyclic wet and dry (XD3)	Wet, rarely dry (XD2)	Moderate humidity (XD1)
Appearance	Red-brown and dense	Yellow-brown and relatively soft	Yellow-brown and relatively soft	Red-brown and very dense
Image				

Samples	E	F	G	H
Source	A steel plate near the sea, Ningbo, China	The rebar in RC port, Yokosuka, Japan	The rebar in RC specimens electro-osmosis-treated	The rebar electrochemically treated in NaCl solution
Collecting method	Knocking	Knocking and peeling	Peeling	Drying suspended products
Conservation method	Packed in bags	Packed in bags	Packed in bags	Packed in bags
Corroding time	6 years	40 years	2 months	1 month
Conserving time	2 years	1 month	1 month	1.5 years
Classical environment[c]	Structures near or on the coast (XS1)	Tidal, splash, and spray zones (XS3)	Wet, rarely dry (XD2)	Permanently submerged (XS2)
Appearance	Dark brown and dense	Dark brown and very dense	Red-brown and relatively dense	Black and very soft
Image				

[a]*Frequently wetted by sewage.*
[b]*Part of this ironwork is submerged in waste water; the rust sample was peeled from the above part exposed to air.*
[c]*As Eurocode 2: Design of concrete structures—Part 1-1: General rules and rules for buildings (EN1992-1-1:2004).*

the rust samples in this study came from chloride-containing environments. Fe_2O_3 was also believed to exist in the rust samples because the individual Fe_2O_3 peaks appeared in the XRD patterns of most rust samples. Fe_2O_3, however, was considered to comprise the mill scale that formed before corrosion initiation, but it was not generated during the rust expansion process because the formation of Fe_2O_3 required a very high temperature [23].

As shown in Fig. 3.1, the peak intensities of these compositions for each sample were different, indicating that the percentage of every composition was different in each rust sample because the rust samples were generated in different environments.

However, it needs to be pointed out that although XRD is a good way to check the ratio between the different products, this technique was not able to differentiate the rust compositions quantitatively in this study. This is because most peaks were marked with a variety of compositions in the XRD diffraction patterns obtained in this test, but the XRD technique could not differentiate between the attribute of each composition for every peak. Therefore, thermal analysis was introduced in this study to quantitatively analyze the compositions of the rust samples, which is explained in the next section.

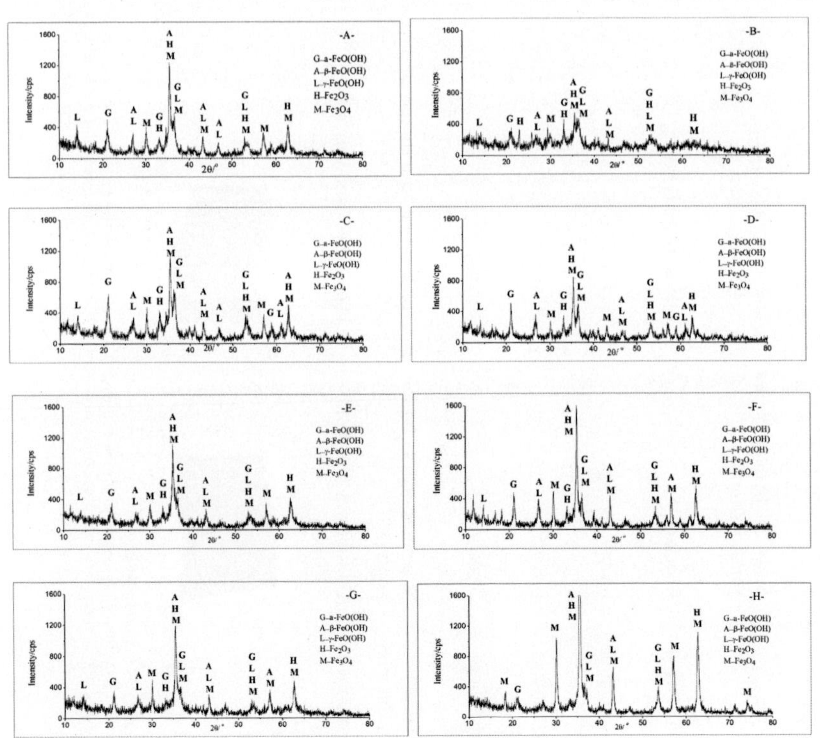

FIGURE 3.1 XRD pattern of eight different rust samples.

It also needs to be noted that more or less amorphous products exist in rust; however, the XRD spectra can only determine the presence of crystals but not whether something is noncrystalline. Amorphous material could not be quantitatively analyzed [26]. Therefore, the amorphous products are not considered in this study.

3.2.2.2 *Thermogravimetric Analysis*

TG curves of all samples are given in Fig. 3.2. Two distinct steps, 20–200°C and 200–400°C, are observed in every single TG curve. Step I, the decline stage from 20°C to 200°C in Fig. 3.2, indicates the dehydration of the rust sample resulting from the evaporation of physically adsorbed water in the rust. Step II, the steep decline stage from 200°C to 400°C, shows that the hydroxyl in the hydroxy-oxide is destroyed [23] as follows:

$$2FeO(OH) \rightarrow Fe_2O_3 + H_2O \tag{3.1}$$

During the heating process, the magnetite is continuously oxidized to γ-Fe_2O3 when being heated after 200°C. However, when the heating temperature is higher than 300°C, the transformation proceeds further to α-Fe_2O3 [23]. The reaction process is as follows:

$$\text{magnetite}(Fe_3O_4) \xrightarrow{<300^\circ C} \text{maghemite}(\gamma - Fe_2O_3) \xrightarrow{>300^\circ C} \text{hematite}(\alpha - Fe_2O_3)$$

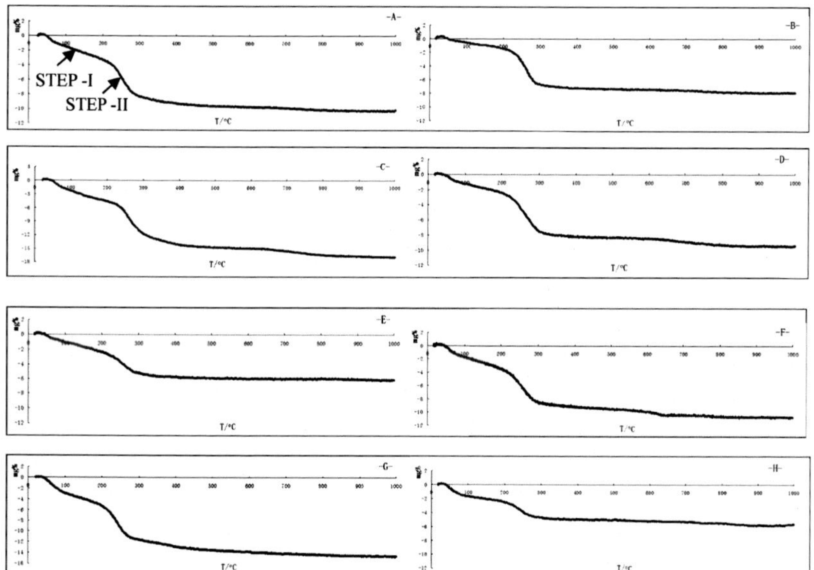

FIGURE 3.2 TG curves of all rust samples.

It can be seen that although different products will be generated during the heating process, Fe_3O_4 will gradually transfer into α-Fe_2O_3 after being heated above 300°C. Therefore, the transformation can be simplified as follows:

$$4Fe_3O_4 + O_2 \rightarrow 6Fe_2O_3 \tag{3.2}$$

The mass increase due to the addition of oxygen in Reaction (2) cannot be shown in TG curves because it is combined with the mass loss in Reaction (1), but this oxidation reaction can be detected by DTA.

Recording the mass change during 200–400°C for each rust sample, the content of the hydroxy-oxide FeO(OH) and the oxidation Fe_3O_4 in the rust sample can be estimated by the ratio of molar mass between FeO(OH) and H_2O in Eq. (3.1) and the ratio of molar mass between Fe_3O_4 and O_2 in Eq. (3.2), which are discussed in Section 3.2.4.1.

3.2.2.3 *Differential Thermal Analysis*

During thermal treatment, the DTA curves showed no significant changes after 600°C. Therefore, the DTA curves are plotted only for room temperature up to 600°C for all samples (see Fig. 3.3). Fig. 3.3 shows that

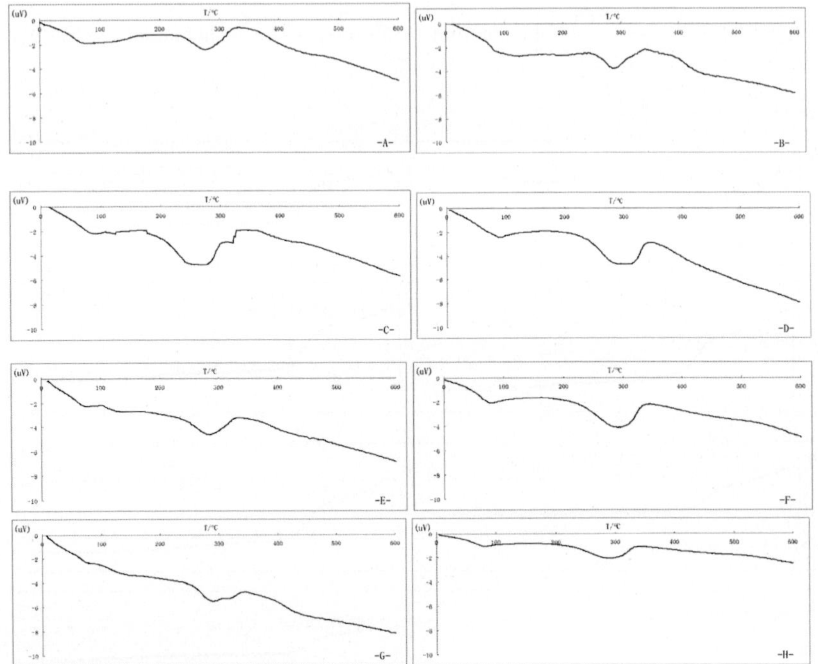

FIGURE 3.3 DTA curves of all rust samples.

there were two endothermic reactions indicated by the curves, one at 20–200°C and the other at 200–400°C, corresponding to the two steps in the TG analysis. Relatively low energy is absorbed during the first endothermic reaction, corresponding to the evaporation of the physically absorbed water from the sample. The deeper concave curve indicates violent absorption of energy during the latter endothermic reaction, which corresponds to the dehydration (ie, the hydroxyl in the hydroxy-oxide is destroyed) [23] as indicated by Eq. (3.1). The more energy that is absorbed during the second endothermic reaction, the greater the percentage of hydroxy-oxide contained in the rust sample. The exothermic peak that follows is due to the oxidation of magnetite as indicated by Eq. (3.2).

3.2.3 Composition of Rust Samples

3.2.3.1 Influencing Factors

The formation of corrosion products is affected by environmental factors [23,27,28], such as humidity, temperature, oxygen supplementation, and harmful ions (ie, Cl^-, SO_4^{2-}). As shown in Fig. 3.1, although all rust samples had similar compositions, the peak intensities of the compounds for each sample were different due to their different environments. The influence of the environment is discussed in terms of the following three aspects.

1. Oxygen supplementation

   Rich oxygen can improve the content of high-quantivalency iron ions, whereas poor oxygen supplementation induces low-quantivalency iron ions. In the absence of oxygen, the corrosion process will cease. Sample H was generated in water, a very poor oxygen supplementation environment; therefore, its color is black, indicating its low-quantivalency iron ions. Sample F was formed in concrete and the oxygen supplementation was not sufficient; therefore, its color is dark brown, as shown in Table 3.1. The other rust samples that were generated in the air contain more high-quantivalency iron ions and therefore are yellow-brown or red-brown.
2. Humidity

   Humidity affects the electrolytic resistance and transport of ions and provides the hydroxyl ions for the production of hydroxy-oxides. Environmental humidity determines the content of hydroxy-oxides. The hydroxy-oxides were identified by XRD in all rust samples. The hydroxy-oxide content of every rust sample could be estimated by thermal analysis, which is discussed in Section 3.2.4.1.

3. Ions

In the environments containing rich chloride ions the content of β-FeO(OH) is usually high; under the influence of SO_4^{2-} and NO_3^-, rust is mainly composed of α-FeO(OH) and γ-FeO(OH), where γ-FeO(OH) slowly transforms to α-FeO(OH) [29]. Sample A, which was from a coastal city, was consequentially affected by salt mist. Samples B and C were in contact with waste water containing chloride ions. In the durability laboratory of Zhejiang University, sample D was affected by salt mist evaporated from tanks of saline solution. Samples E and F were collected from structures by the sea. The steel bars in samples G and F were treated with electrochemical techniques using saline solution as the medium for current transfer to accelerate the corrosion process. Therefore, β-FeOOH was found in all rust samples and was the dominant hydroxyl-oxide.

3.2.3.2 Improving Rust Composition Analysis Using the TG Technique

The hydroxy-oxides in the rust sample transferred to the oxides [23] during the heating process. Comparing the XRD patterns of the original and the heated samples improved the rust composition analysis. Sample A is used as an example here to present the analysis process. Fig. 3.4 shows XRD patterns of both the original and the heated sample A. Twelve peaks in the original XRD pattern of sample A are marked as I, II, III, IV, V, VI, VII, VIII, IX, X, XI, and XII.

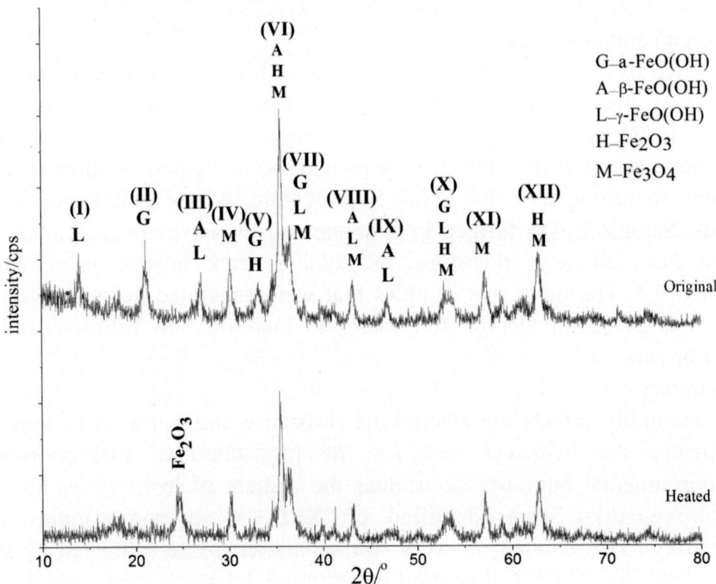

FIGURE 3.4 XRD patterns of the original and heated sample 1.

The following can be found from comparisons of these two patterns:

1. The diffraction peaks of hydroxy-oxides I, II, III, and IX all disappeared after combustion.
2. A new peak occurred at 25° after combustion, corresponding to Fe_2O_3, which means that the hydroxy-oxides were transferred to the oxide Fe_2O_3 as indicated by Eq. (3.1).
3. The peaks of VII, VIII, X, and XII did not change during thermal treatment. Therefore, these peaks correspond to Fe_2O_3 or Fe_3O_4 only and not to hydroxy-oxides.
4. Peak V disappeared after combustion, which indicates that this peak only corresponds to α-FeO(OH), not Fe_2O_3.
5. For peak VI, the descending height reflected the dehydration of β-FeO(OH).
6. Due to its stability in the heating process, peaks IV and XI showed no changes and thus indicate Fe_3O_4.

According to the aforementioned analysis, the improved XRD pattern for sample A is shown in Fig. 3.5. Comparing the XRD pattern of sample A in

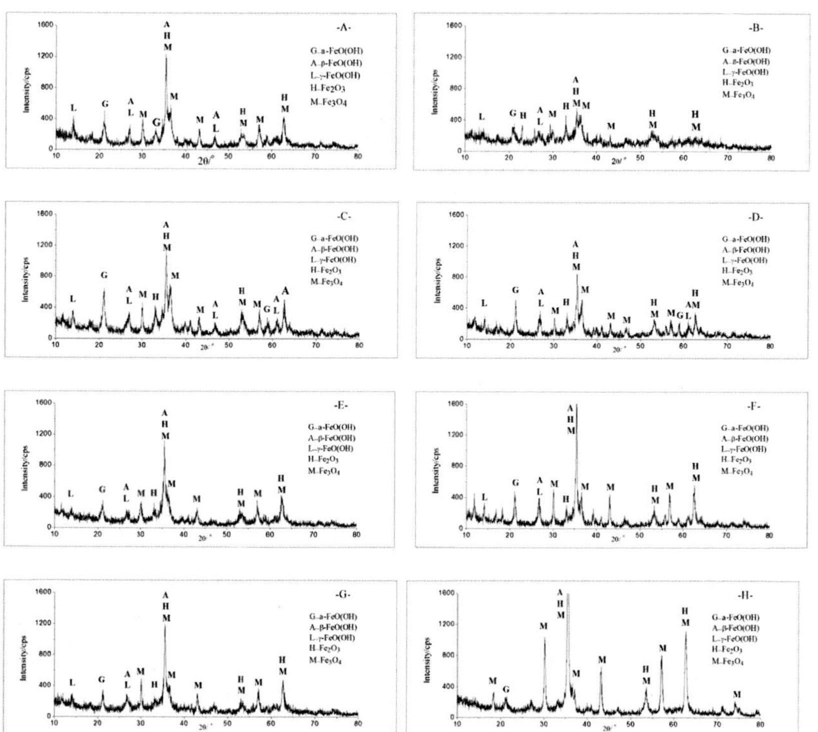

FIGURE 3.5 Improved XRD patterns of eight rust samples.

Fig. 3.1 and in Fig. 3.5, we can see improvement in the identification of some peaks. This is also true of the XRD spectra of the other samples. The improved XRD patterns for the other samples are also shown in the same figure.

It should be pointed out that sample H was heated at 60°C to dry the rust suspended in the saline solution before XRD and thermal analysis, which is probably why the content of hydroxyl-oxides is not significant even though this sample had once contacted water directly.

3.2.4 Expansion Coefficient of Rust Samples

3.2.4.1 Classification of Hydroxy-oxides and Oxides

Based on the water reduction in the range of 200–400°C of the TG curves, all samples can be divided into two categories: the oxygen oxides and the hydroxy-oxides. The contents of these two categories of products can be calculated based on Eqs. (3.1) and (3.2). Table 3.2 lists the results.

3.2.4.2 Types of Hydroxy-oxides and Oxides

There are different types of hydroxy-oxides, namely α-FeO(OH), β-FeO (OH), and γ-FeO(OH), and oxides, namely Fe_2O_3 and Fe_3O_4. γ-FeO(OH) is an unstable composition that can transform into α-FeO(OH) slowly in normal conditions without ion ingression [29]. For more information about transformation of steel corrosion products, please refer to Section 2.4 "Corrosion Products" in chapter "Steel corrosion in concrete".

TABLE 3.2 The Content of Two Categories: Products for Eight Samples (mg)

Sample	Total Mass	Mass Loss During 0–200°C	Mass Loss During 200–400°C	Mass of Hydroxy-oxide	Mass of Oxides
A	9.41	0.38	0.49 (5.26%)	5.69	3.34
B	9.66	0.17	0.56 (5.77%)	6.23	3.26
C	9.38	0.42	0.89 (9.53%)	8.83	0.13
D	8.75	0.22	0.51 (5.78%)	5.69	2.84
E	9.34	0.23	0.31 (3.32%)	4.17	4.94
F	10.01	0.38	0.56 (5.62%)	6.30	3.33
G	10.00	0.55	0.71 (7.09%)	7.47	1.98
H	9.39	0.26	0.23 (2.41%)	3.63	5.50

Based on the XRD/TG/DTA analyses in this study and the environments in which the rust samples were produced, the types of hydroxy-oxides and oxides of the rust samples are discussed here.

1. Hydroxy-oxides

   According to Fig. 3.5, all rust samples contain more β-FeO(OH) compared with the other two kinds of hydroxy-oxide due to the influence of the chloride ions in the environments [23]. The percentages of three hydroxyl-oxides can be estimated by their peak values and the width of each peak in Fig. 3.5.
2. Oxides

   Fe_2O_3 and Fe_3O_4 were detected in all rust samples. As has been pointed out before, Fe_2O_3 cannot be a part of corrosion products because the formation of Fe_2O_3 requires a high temperature [23]. Therefore, Fe_2O_3 does not contribute to the volume expansion acting on the surrounding concrete cover and the subsequent calculations of the expansion coefficients of rust do not consider Fe_2O_3.

3.2.4.3 *Expansion Coefficients for Rust Samples*

The rust expansion coefficient can be calculated as listed in Table 3.3.

Based on the content of the hydroxy-oxides and the oxides listed in Table 3.2, the rust expansion coefficients of eight samples could be calculated and the results are given in Table 3.4. It is clear that the rust samples have different expansion coefficients, varying from 2.64 to 3.24, due to their different environments. This enables us to establish the relationship between the environments and the expansion coefficients of rust.

TABLE 3.3 The Expansion Coefficient of the Main Hydroxy-Oxides and Oxides

Corrosion Product	Expansion Coefficient
Fe_2O_3	2.15
Fe_3O_4	2.10
α-FeO(OH)	2.95
β-FeO(OH)	3.53
γ-FeOOH	3.07

TABLE 3.4 The Expansion Coefficients for All Rust Samples

Samples	Rust Expansion Coefficient
A	2.94
B	2.96
C	3.24
D	2.98
E	2.85
F	3.02
G	3.14
H	2.64

TABLE 3.5 Exposure Classes Related to Environmental Conditions

Factor	Level	Environmental Condition
Humidity	High	Subjected to long-term water contact or submerged in water
	Middle	Subjected to direct water contact occasionally (tidal zone and splash zone)
	Low	Normal atmosphere
Oxygen supplement	High	Directly exposed to air
	Middle	Insufficient supplementation of oxygen (the rebar surface after the cover cracking)
	Low	Oxygen deficiency (parts of the member and the rebar surface were immersed in water before the cover began cracking)

3.2.4.4 *Rust Expansion Coefficients in Different Environments*

As stated in Section 3.2.3.1, humidity and oxygen supplementation are the two main influencing factors. Three levels classified by the specific environmental conditions are proposed in Table 3.5 with regard to these two factors.

According to Fig. 3.5 and Table 3.5, the environmental conditions for all samples can be described and classified as shown in Table 3.6.

TABLE 3.6 Environmental Classifications for All Rust Samples

Samples	Oxygen Supplement	Humidity	Appearance of Cl^-	Description
A	High	Low	Yes	Formed freely in a coastal city atmosphere
B	High	Middle	Yes	Formed freely in an atmosphere affected by the waste water
C	High	High	Yes	Formed freely in an atmosphere affected by waste water
D	High	Low	Yes	Formed freely in an atmosphere with saline mist
E	High	Low	Yes	Formed freely in an atmosphere beside the sea
F	Middle	Middle	Yes	Formed in a concrete member beside the sea
G	Low	High	Yes	Formed in a concrete member treated with electro-osmosis
H	Low	Low	Yes	Formed from a steel bar electrochemically treated in the saline solution and heated at 60°C for drying

Based on the results shown in Fig. 3.5, Tables 3.4 and 3.6, further discussion of the relationship between the environments and the rust expansion coefficients are stated here.

1. For the high-humidity and high-oxygen supplementation (HH-HO) level, such as sample C, the rust expansion coefficient is near 3.3; however, this parameter is approximately 3.1 for the high-humidity and low-oxygen supplementation (HH-LO) level, such as sample G. The rust expansion coefficient of the high-humidity and low-oxygen supplement (HH-MO) level is thus estimated by averaging these two values, that is, 3.2.
2. For the low-humidity and high-oxygen supplementation (LH-HO) level, such as sample A, the expansion coefficient is approximately 2.9; however, this parameter is approximately 2.6 for the low-humidity and low-oxygen supplementation (LH-LO) level, such as sample H. For the

low-humidity and the middle oxygen supplementation (LH-MO) level, an average of these two can be used.

3. For the middle humidity and high-oxygen supplementation (MH-HO) level, the expansion coefficient is approximately 3.1 by averaging the values of the HH-HO level and the LH-HO level. This parameter is approximately 3.0 for the middle humidity and middle oxygen supplementation (MH-MO) level, such as sample F. The rust expansion coefficient of the middle humidity and low-oxygen supplementation (MH-LO) level can be estimated by the linear extrapolation of the values of the MH-HO level and the MH-MO level (ie, 2.9), which is reasonable given the expansion coefficients of HH-LO and LH-LO.

Based on this discussion, the rust expansion coefficients corresponding to various environments are proposed, as given in Table 3.7. These values can be utilized in the analysis of the concrete cover cracking due to reinforcement corrosion. It should be noted that the oxygen supplementation for the steel bar in concrete structures before cracking is "low" due to the protection of the concrete cover. It also needs be pointed out that all the rust expansion coefficients given in Table 3.7 consider the influence of chloride ions because all rust samples came from environments containing chloride ions. For samples without chloride ions, 0.5 should be subtracted from the rust expansion coefficients in Table 3.7 for application because the expansion coefficient of β-FeO(OH) is 3.5, which is nearly 0.5 larger than that of α-FeO(OH) or γ-FeO(OH).

3.3 MODULUS OF STEEL CORROSION PRODUCTS IN CONCRETE

3.3.1 Experimental Program

Section 3.2 has showed that the corrosion products obtained in the laboratory differed in component content compared to natural corrosion products. Therefore, to obtain the actual rust modulus to be used in the model of concrete cracking because of steel corrosion, the rust used in this study was

TABLE 3.7 Rust Expansion Coefficients Corresponding to Different Environments

Environmental Factors		Humidity (Water)		
		High	Middle	Low
Oxygen supplement	High	3.3	3.1	2.9
	Middle	3.2	3.0	2.8
	Low	3.1	2.9	2.6

Note: Subtracting 0.5 for the environment without chloride ions ingression.

peeled from a corroded steel bar that was embedded in a concrete beam. This beam was taken from a concrete port that has been active for approximately 40 years in Yokosuka, Japan. The annual average temperature in this region is approximately 26°C, and the annual average relative humidity is approximately 70%. This beam was in the tidal zone and could be classified as being in the XS3 category according to the service environment classification of EN1992-1-1:2004. The rust was dense and dark brown in color when peeled from the corroded beam, as shown in Fig. 3.6.

After manually peeling from the steel bar, the steel corrosion products were sealed in plastic bags for 1 month before being tested to keep the corrosion products in the same condition as they were in the beam.

Rust was processed into the flaky samples (Fig. 3.6d). The steel corrosion products were ground on both the top and bottom surfaces using different grades of sandpaper. In total, 54 flaky rust samples were prepared and were numbered from Flaky 1 to Flaky 54.

A cyclic low-compression test was performed on the flaky rust samples to study the modulus of steel corrosion products that is used in the concrete cracking model. A CMT4204 universal tester made by SANS Testing Machine was used during the test. The procedures were as follows:

1. Trace the profile of the tested flaky sample on correlate paper, and calculate the area of the flaky sample (defined as A);
2. Measure the height of the tested flaky sample using a vernier caliper three times, and average three heights to represent the height of the flaky sample (defined as ξ);
3. Place the flaky rust sample on the tray of the universal tester, and then load and unload the rust flakes perpendicularly. Next, record the stress–strain data.

FIGURE 3.6 The concrete port and the steel corrosion products. (a) The concrete port in Yokosuka. (b) The corroded steel bar in the concrete beam. (c) The corrosion product peeled from the corroded steel bar. (d) Flaky rust samples.

A stress level from 0 to 12 MPa was chosen for the loading and unloading regions because the pressure induced by the volume expansion of steel corrosion products at the concrete and steel interface was considered to be approximately several MPa [10, 18–21].

3.3.2 Loading and Unloading Stress–Strain Curve

Fig. 3.7 shows the loading and unloading stress–strain curve for one flaky rust sample that was randomly chosen from a total of 54 flaky samples (σ_r and ε_r are the stress and the stain of the flaky rust sample, respectively). The solid lines describe the loading process and the dashed lines denote the unloading process. Clearly, the curve under the first loading has a nonlinear relationship between the stress and the strain, whereas the stress–strain curve at the third cycle shows linear behavior. The slope of the third loading line from 1 to 10 MPa is calculated as the elastic modulus of the flaky rust sample, which is defined as $E_{r,o}$.

The stress–strain curve under the first loading in Fig. 3.7 is similar to the stress–strain curve recommended by Lundgren [16]. Therefore, a power function is used to represent the stress–strain relationship as:

$$\sigma_r = k \cdot \varepsilon_r^p \tag{3.3}$$

where σ_r and ε_r are the stress and the stain of the flaky rust samples, and k and p are the material constants, which can be determined by using the best fit of the equation for the experimental curve shown in Fig. 3.7. Based on a regression analysis of the flaky rust sample, a constitutive model for the rust samples is proposed as:

$$\sigma_r = 0.45\varepsilon_r^{3.9} \tag{3.4}$$

where the unit of σ_r is GPa.

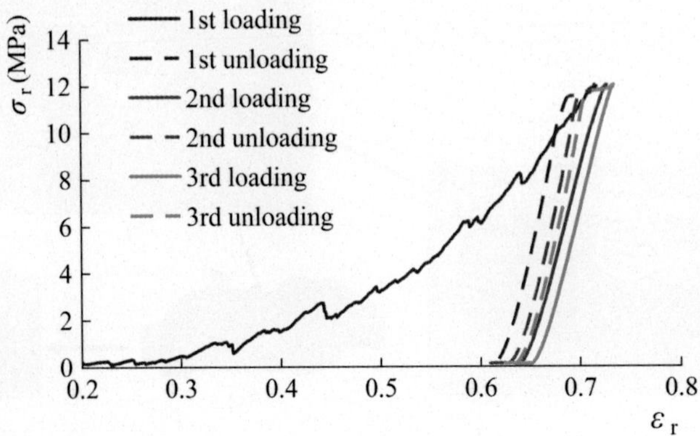

FIGURE 3.7 Typical loading and unloading stress–strain curve.

It should be mentioned that this result was obtained from rust growth on a concrete beam from Yokosuka, Japan. One should be cautious when using this result to describe the stress–strain relationship of rust generated in other areas.

3.3.3 Tested Data of Cyclic Low-Compression Test

The tested data of the samples from Flaky 1 to Flaky 54 are listed in Table 3.8. The height and area of the rust flakes were measured as stated in Section 3.3.1, and the modulus of rust $E_{r,o}$ was calculated using the method stated in Section 3.3.2. The volume and height-to-area ratios ($\psi = \xi/\sqrt{A}$) that were calculated for each flaky rust sample are also listed in Table 3.8.

Based on Table 3.8, the relationships of the rust modulus against the height, area, volume, and the height-to-area ratio are analyzed.

It is found that there is no clear relationship between the height and the modulus. The rust modulus decreased with the increase of both volume and area. However, the tendency of the rust modulus to vary with the rust area is more distinct; therefore, the rust area is considered to be a key parameter. The effect of the rust area on the rust modulus is discussed in Section 3.3.4.

That rust modulus was determined based on geometric features here, and it appears to conflict with the intuitive notion that the modulus is an intrinsic property of rust. This phenomenon is considered to be caused by friction occurring between the rust samples and the loading device, which artificially increase the apparent strength of the rust if the height-to-width ratio of the specimen is too low. Therefore, the term "apparent modulus," which was considered to be more rigorous than "modulus," was proposed to represent the measured values here to highlight the difficulty in measuring an intrinsic modulus in rust samples with complex shapes.

3.3.4 Modulus of Rust

Based on the data in Table 3.8, the relationship between the area A (mm^2) and the modulus $E_{r,o}$ (MPa) can be described as follows:

$$E_{r,o} = \begin{cases} 3760A^{-0.5306} & A < 360\ \text{mm}^2 \\ 165 & A \geq 360\ \text{mm}^2 \end{cases} \tag{3.5}$$

Eq. (3.5) describes the corresponding change of the modulus to different areas. The first equation describes the decreasing tendency of the rust modulus with an increase in area, which means that rust areas have an influence on the rust modulus when the rust size is small. This finding may explain why there are enormous differences in the rust modulus in the literature.

The second equation shows that the rust modulus converges to 165 MPa when the area is greater than 360 mm^2. Because the area of rust in real structures is certainly much larger than 360 mm^2, 165 MPa can be regarded as the modulus of corrosion products in actual projects. It should be noted

TABLE 3.8 Tested Data from the Cyclic Low-Compression Test

Flaky Sample No.	**1**	**2**	**3**	**4**	**5**	**6**	**7**	**8**	**9**
Height, mm	1.17	1.38	2.82	2.56	2.46	2.14	2.44	2.46	2.9
Area, mm^2	45	53	33	42	51	59	75	85	73
Volume, mm^3	53	73	93	108	125	126	183	209	212
Height-to-area ratio	0.174	0.190	0.491	0.395	0.344	0.279	0.282	0.267	0.339
Modulus, MPa	347	256	560	394	330	319	421	352	339
Flaky Sample No.	**10**	**11**	**12**	**13**	**14**	**15**	**16**	**17**	**18**
Height, mm	2.46	1.72	2.3	2.16	3.04	3.03	2.02	2.6	3.52
Area, mm^2	107	163	124	133	118	126	196	166	140
Volume, mm^3	263	280	285	287	359	382	396	432	493
Height-to-area ratio	0.238	0.135	0.207	0.187	0.280	0.270	0.144	0.202	0.297
Modulus, MPa	266	166	227	236	304	319	204	229	332
Flaky Sample No.	**19**	**20**	**21**	**22**	**23**	**24**	**25**	**26**	**27**
Height, mm	4.48	2.7	3.7	2.7	3	4.12	5.03	3.22	3.99
Area, mm^2	119	219	170	245	248	181	155	266	225
Volume, mm^3	533	591	629	662	744	746	780	857	898
Height-to-area ratio	0.411	0.182	0.284	0.172	0.191	0.306	0.404	0.197	0.266
Modulus, MPa	454	161	144	168	128	135	430	147	129
Flaky Sample No.	**28**	**29**	**30**	**31**	**32**	**33**	**34**	**35**	**36**
Height, mm	6.06	4.11	3.9	7.78	7.34	5.09	4.87	5.14	2.64
Area, mm^2	158	253	355	177	242	351	380	411	124
Volume, mm^3	957	1040	1384	1377	1776	1787	1851	2113	327
Height-to-area ratio	0.482	0.258	0.207	0.585	0.472	0.272	0.250	0.254	0.237
Modulus, MPa	513	109	96	576	475	95	95	91	298

(Continued)

TABLE 3.8 (Continued)

Flaky Sample No.	37	38	39	40	41	42	43	44	45
Height, mm	5.25	3.58	5	5.15	4.88	4.23	3.41	3.63	4.28
Area, mm^2	139	170	222	297	428	87	122	143	188
Volume, mm^3	730	609	1110	1530	2089	368	416	519	805
Height-to-area ratio	0.445	0.275	0.336	0.299	0.236	0.454	0.309	0.304	0.312
Modulus, MPa	440	260	356	291	204	576	422	345	335
Flaky Sample No.	**46**	**47**	**48**	**49**	**50**	**51**	**52**	**53**	**54**
Height, mm	4.86	2.87	4.38	5.12	5.05	7.06	5.41	3.33	6.15
Area, mm^2	380	74	90	153	116	408	430	505	579
Volume, mm^3	1847	212	394	1089	586	2880	2326	1682	3561
Height-to-area ratio	0.249	0.334	0.462	0.576	0.469	0.350	0.261	0.148	0.256
Modulus, MPa	227	519	543	179	610	311	207	117	172

that 165 MPa specifically describes the modulus of the rust collected for the concrete port in Yokosuka, Japan. Although the value of the rust modulus may be affected by the sample selection, it can be proposed that, based on this study, the order of magnitude of the modulus is 100 MPa. This result agrees with a previous study [5] that found that the modulus of corrosion products was in the range of 0.1–0.5 GPa. Therefore, 100 MPa is suggested to be the magnitude of the elastic modulus of steel corrosion products and can be used in the model of corrosion cracking induced by steel corrosion.

3.4 CONCLUSIONS

1. The expansion coefficients of eight rust samples from different environments were estimated by the weighted-average method based on the percentage of the compositions in each sample.
2. The rust expansion coefficients corresponding to the different environments were proposed using humidity and oxygen supplementation as the most influential factors. These can be applied to the analysis of the concrete cover cracking due to reinforcement corrosion.

3. The cyclic low-compression tests found that the modulus of steel corrosion products decreased with the increase of the sample areas when they were less than 360 mm^2, whereas the modulus tended to be constant when the areas were larger than 360 mm^2.
4. A value of 165 MPa was found, in this work, to specifically describe the apparent modulus of the rust. We suggest 100 MPa to be the magnitude of the apparent modulus for steel corrosion products that can be used in the model of concrete cracking induced by steel corrosion.

REFERENCES

[1] Tuutti K. Corrosion of steel in concrete. Report 4. Stockholm, Sweden: Swedish Cement and Concrete Research Institute; April 1982. p. 82.

[2] Molina FJ, Alonso C, Andrade C. Cover cracking as a function of rebar corrosion: part 2—numerical model. Mater Struct 1993;26(9):532–48.

[3] Noghabai K. Effect of tension softening on the performance of concrete structures: experimental, analytical and computational studies, 1997. Lulea University of Technology; Lulea, Sweden.

[4] Lundgren K. Modelling the effect of corrosion on bond in reinforced concrete. Mag Concrete Res 2002;54(3):165–73.

[5] Wang HL, Jin WL, Sun XY. Fracture model for protective layer cracking of reinforced concrete structure due to rebar corrosion. J Hydraulic Eng 2008;39(4):863–9 [in Chinese].

[6] Lu CH, Zhao YX, Jin WL. Modeling of time to corrosion-induced cover cracking in reinforced concrete structures. J Build Struct 2010;31(2):85–92.

[7] Zhao YX, Jin WL. Modeling the amount of steel corrosion at the cracking of concrete cover. Adv Struct Eng 2006;9(5):687–96.

[8] Bhargava K, Ghosh AK, Mori Y, et al. Analytical model for time to cover cracking in RC structures due to rebar corrosion. Nucl Eng Design 2006;236(11):1123–39.

[9] Val DV, Chernin L, Stewart MG. Experimental and numerical investigation of corrosion-induced cover cracking in reinforced concrete structures. J Struct Eng 2009;135(4):376–85.

[10] Liu Y, Weyers RE. Modeling the time-to-corrosion cracking in chloride contaminated reinforced concrete structures. ACI Mater J 1998;95(6):675–81.

[11] Caré S, Nguyen QT, L'Hostis V, et al. Mechanical properties of the rust layer induced by impressed current method in reinforced mortar. Cement Concrete Res 2008;38(8):1079–91.

[12] Syed S. Atmospheric corrosion of hot and cold rolled carbon steel under field exposure in Saudi Arabia. Corrosion Sci 2008;50(6):1779–84.

[13] Samsonov GV. The oxide handbook. New York, NY: Ifi/Plenum; 1973.

[14] Suda K, Misra S, Motohashi K. Corrosion products of reinforcing bars embedded in concrete. Corrosion Sci 1993;35(5):1543–9.

[15] Petre-Lazar I, Gerard B. Mechanical behaviour of corrosion products formed at the steel-concrete interface. Testing and modelling. Proceedings of EM2000, 14th Engineering Mechanics Conference, Austin, Texas, USA. 2000.

[16] Lundgren K. Bond between ribbed bars and concrete. Part 2: the effect of corrosion. Mag Concrete Res 2005;57(7):383–96.

[17] Ouglova A, Berthaud Y, François M, et al. Mechanical properties of an iron oxide formed by corrosion in reinforced concrete structures. Corrosion Sci 2006;48(12):3988–4000.

[18] Andrade C, Alonso C, Molina FJ. Cover cracking as a function of bar corrosion: part I-experimental test. Mater Struct 1993;26(8):453–64.

[19] Rasheeduzzafar, Al-Saadoun SS, Al-Gahtani AS. Corrosion cracking in relation to bar diameter, cover, and concrete quality. J Mater Civil Eng 1992;4(4):327–42.

[20] Zhao YX, Jin WL. Numerical-based method for calculating reinforcement corrosion at concrete cover cracking due to corrosion. J Zhejiang Univ (Engineering Science) 2008;6:36.

[21] Bazant ZP. Physical model for steel corrosion in concrete sea structures–application. J Struct Div 1979;105 (ASCE 14652 Proceeding).

[22] Chen J, Li C. Thermal analysis and application. Beijing: Science Press; 1985 [in Chinese].

[23] Cornell RM, Schwertmann U. The iron oxides: structure, properties, reactions, occurrences and uses. Weinheim, Germany: John Wiley & Sons; 2006.

[24] Zitrou E, Nikolaou J, Tsakiridis PE, et al. Atmospheric corrosion of steel reinforcing bars produced by various manufacturing processes. Constr Building Mater 2007;21(6):1161–9.

[25] De la Fuente D, Díaz I, Simancas J, et al. Long-term atmospheric corrosion of mild steel. Corrosion Sci 2011;53(2):604–17.

[26] Asami K, Kikuchi M. In-depth distribution of rusts on a plain carbon steel and weathering steels exposed to coastal–industrial atmosphere for 17 years. Corrosion Sci 2003;45(11):2671–88.

[27] Li QX, Wang ZY, Han W, et al. Characterization of the rust formed on weathering steel exposed to Qinghai salt lake atmosphere. Corrosion Sci 2008;50(2):365–71.

[28] Yamashita M, Konishi H, Kozakura T, et al. In situ observation of initial rust formation process on carbon steel under Na_2SO_4 and NaCl solution films with wet/dry cycles using synchrotron radiation X-rays. Corrosion Sci 2005;47(10):2492–8.

[29] Ishikawa T, Takeuchi K, Kandori K, et al. Transformation of γ-FeOOH to α-FeOOH in acidic solutions containing metal ions. Colloids Surf A Physicochem Eng Asp 2005;266 (1):155–9.

Chapter 4

Damage Analysis and Cracking Model of Reinforced Concrete Structures with Rebar Corrosion

Chapter Outline

4.1 INTRODUCTION

Considerable analytical research has been undertaken regarding the corrosion-induced cracking process and is reviewed in chapter "Introduction." As has been discussed in Section 1.3.3, some of these models [1–7] were based on the theory of elasticity and did not consider the residual strength of cracked concrete; also, most of the existing models

Steel Corrosion-Induced Concrete Cracking. DOI: http://dx.doi.org/10.1016/B978-0-12-809197-5.00004-9

[1,2,4−6,8−11] neglected the behavior of the rust layer between the steel bar and concrete cover.

In this chapter, damage mechanics and elastic mechanics are utilized to investigate the stresses and strains in the surrounding concrete during the corrosion-induced cracking process and the amount of steel corrosion at the cracking of the concrete cover by taking into account the mechanical properties of both uncracked and cracked concrete as well as the rust products. Expansive pressure and steel corrosion were analyzed. Parametric studies were performed to discuss the effects of the correlative factors.

4.2 BASIC CONCRETE CRACKING MODEL DUE TO STEEL CORROSION

In this chapter, our focus is on the period from stress initiation to the cracking of the concrete surface. In common with the previous cracking models [1−11], concrete with an embedded reinforcing steel bar is modeled as a thick-walled cylinder with a wall thickness equal to the thinnest concrete cover. The cylinder is assumed to be subject to internal pressure due to the formation of corrosion products. The corrosion products are assumed to be uniformly formed around the steel surface, and the uniform corrosion products result in uniform internal expansive pressure. This is rarely the case in practice; nonuniform corrosion around the rebar perimeter is usually the reality. However, it can significantly simplify the modeling and help us to understand the corrosion-induced cracking process. The nonuniform corrosion scenario is discussed later in chapters "Nonuniform Distribution of Rust Layer Around Steel Bar in Concrete" and "Corrosion-Induced Concrete Cracking Model Considering Corrosion Products-Filled Paste."

Fig. 4.1 illustrates the interactions of steel bar, corrosion products, and the surrounding concrete. The damage process of the concrete cover can be divided into two distinct stages: the noncracking stage and the partial cracking stage, as illustrated in Fig. 4.1. In Fig. 4.1, C indicates the thickness of the concrete cover, d is the diameter of steel bar, R is the radius of the steel bar, R_c is the radius at the interface between the cracked and intact cylinders, q_{R_c} is the radius pressure at the interface between the intact and cracked cylinders, q is the radius expansive pressure induced by steel corrosion at the steel−concrete interface, δ_c is the deformation of concrete at the steel−concrete interface, δ_r is the deformation of corrosion products at the concrete−rust interface, and d_ρ is the residual diameter of the steel bar after corrosion. Fig. 4.1a shows the model in the noncracking stage. During this stage, elastic mechanics are utilized to analyze the whole concrete cover subjected to corrosion expansion. Once the inner cracks are generated between the concrete and steel, the partial cracking stage begins. Fig. 4.1b shows the analysis of the partial cracking stage. The concrete cover is divided into two

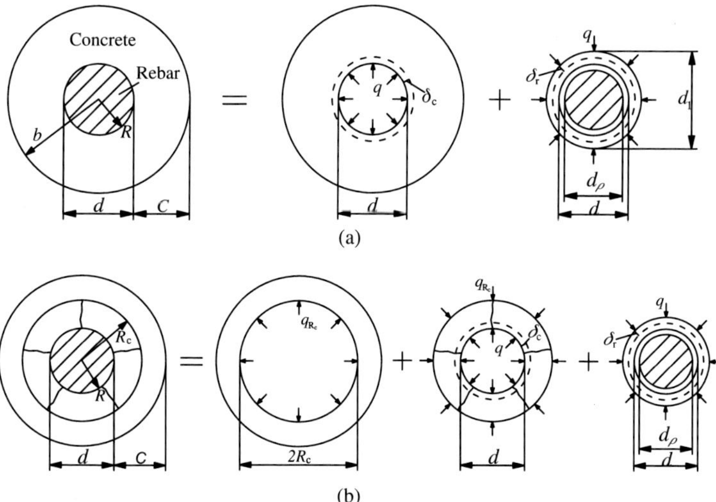

FIGURE 4.1 Deformations of the rust layer and surrounding concrete under expansive pressure. (a) Noncracking. (b) Partial cracking.

areas at this stage. One is the cracked part and the other is the intact part. Elastic mechanics are used to analyze the intact part and damage mechanics are applied to deal with the cracked part. A damage variable, based on Mohr–Coulomb failure criterion [12] and the Mazar damage model [13], is adopted to describe the different damage along the radial direction of the concrete cover. This is discussed in detail in Section 4.4.2.

4.3 NONCRACKING STAGE OF CORROSION-INDUCED CONCRETE CRACKING PROCESS

1. Stress, strain, and displacement

   Fig. 4.1a shows the schematic diagrams used in the analyses. According to elastic mechanics, the stresses of a thick-walled cylinder subjected to internal radial pressure are expressed as follows [14]:

$$\begin{cases} \sigma_r = \dfrac{qR^2}{b^2 - R^2}\left(1 - \dfrac{b^2}{r^2}\right) \\ \sigma_\theta = \dfrac{qR^2}{b^2 - R^2}\left(1 + \dfrac{b^2}{r^2}\right) \end{cases} \tag{4.1}$$

where $b = R + C$, σ_r, and σ_θ are the radial and hoop stresses induced by steel corrosion in concrete cover, respectively. The strains are also expressed as [14]:

$$\begin{cases} \varepsilon_r^e = \dfrac{1+\nu_c}{E_c}\dfrac{qR^2}{b^2-R^2}\left(1-2\nu_c-\dfrac{b^2}{r^2}\right) \\ \varepsilon_\theta^e = \dfrac{1+\nu_c}{E_c}\dfrac{qR^2}{b^2-R^2}\left(1-2\nu_c+\dfrac{b^2}{r^2}\right) \end{cases} \tag{4.2}$$

where E_c is the elastic modulus of concrete, ν_c is the Poisson's ratio of concrete, $\nu_c = 0.3$ in the study as it tends to get larger at a higher stress level [15], and ε_r^e and ε_θ^e are the radial and hoop elastic strains induced by steel corrosion in the concrete cover, respectively.

The concrete cylinder is assumed to crack when the hoop tensile strain at the interface between the steel and the concrete has reached the ultimate tensile strain of concrete. Because there has been plastic strain in concrete before it reaches ultimate tensile stress, the elastic modulus E_c in Eq. (4.2) should be replaced by the deformation modulus at the maximum tensile stress E_c' when calculating the real strain at the inner surface cracking moment, which could be expressed as:

$$\begin{cases} \varepsilon_r = \dfrac{1+\nu_c}{E_c'}\dfrac{qR^2}{b^2-R^2}\left(1-2\nu_c-\dfrac{b^2}{r^2}\right) \\ \varepsilon_\theta = \dfrac{1+\nu_c}{E_c'}\dfrac{qR^2}{b^2-R^2}\left(1-2\nu_c+\dfrac{b^2}{r^2}\right) \end{cases} \tag{4.3}$$

The deformation modulus at the maximum tensile stress $E_c' = 0.5E_c$ [16]. Because $\varepsilon_r = \frac{\partial u_r}{\partial r}$, the deformation could be expressed as:

$$u_r = \frac{1+\nu_c}{E_c'}\frac{qR^2}{b^2-R^2}\left(1-2\nu_c+\frac{b^2}{R^2}\right)r \tag{4.4}$$

where ε_r and ε_θ are the radial and hoop strains induced by steel corrosion in concrete cover, respectively, and u_r is the radial deformation.

At the concrete inner surface cracking moment,

$$\varepsilon_\theta|_{r=R} = \varepsilon_t \tag{4.5}$$

where ε_t is the ultimate tensile strain of concrete, which can be calculated by $\varepsilon_t = f_t/E_c'$, where f_t is the ultimate tensile strength of concrete.

The expansive pressure q at the inner cracking can be obtained by substituting Eq. (4.5) into Eq. (4.6):

$$q^{\text{inner}} = \frac{E'_c}{1+\nu_c} \cdot \frac{\varepsilon_t(b^2 - R^2)}{R^2} \cdot \frac{1}{1 - 2\nu_c + b^2/R^2} \tag{4.6}$$

2. Steel corrosion

The deformation of concrete at the interface between the steel and the concrete is equal to u_r:

$$\delta_c = u_r = \frac{1+\nu_c}{E'_c} \frac{qR^2}{b^2 - R^2} \left(1 - 2\nu_c + \frac{b^2}{R^2}\right) R \tag{4.7}$$

The deformation compatibility of concrete and corrosion products and the deformation of corrosion products can be expressed as follows:

$$R + \delta_c = R_1 + \delta_r, \quad \delta_r = -\frac{n(1-\nu_r^2)R \cdot \sqrt{(n-1)\rho + 1}}{E_r\{[(1+\nu_r)n - 2] + 2/\rho\}} \cdot q \tag{4.8}$$

where R_1 is the nominal radius of the steel bar with free-expansion corrosion products, $R_1 = R\sqrt{(n-1)\rho + 1}$, and E_r and ν_r are the elastic modulus and Poisson ratio of the corrosion products. As proposed in chapter "The Expansion Coefficients and Modulus of Steel Corrosion," E_r is taken as 10^2 MPa, $\nu_r = 0.25$, n is the ratio between the volumes of corrosion products and basic steel, which can be seen in Table 3.7 according to various environments, and ρ is the steel corrosion.

Using Eqs. (4.7) and (4.8), the expansive pressure q can be expressed as follows:

$$q = \frac{\sqrt{(n-1)\rho + 1} - 1}{\frac{(1+\nu_c)(R+C)^2 + (1-\nu_c)R^2}{E'_c(2R \cdot C + C^2)} + \frac{n(1-\nu_r^2) \cdot \sqrt{(n-1)\rho + 1}}{E_r\{[(1+\nu_r)n - 2] + 2/\rho\}}} \tag{4.9}$$

Substituting $x = \sqrt{(n-1)\rho + 1}$ into Eq. (4.9), yields:

$$\alpha_1 x^3 + \alpha_2 x^2 + \alpha_3 x + \alpha_4 = 0 \tag{4.10}$$

let $M_1 = (1+\nu_c)(R+C)^2 + (1-\nu_c)R^2/E'_c(2RC + C^2)$, $M_2 = n(1-\nu_r^2)/E_r$, and $M_3 = (1+\nu_r)n - 2$, and then we have the following expressions, $\alpha_1 = M_2 q - M_3$, $\alpha_2 = M_3 + qM_1M_3$, $\alpha_3 = M_3 - 2(n-1) - M_2 q$, and $\alpha_4 = 2qM_1(n-1) + 2(n-1) - qM_1M_3 - M_3$.

The solution of Eq. (4.9) can be expressed as follows:

$$x = \sqrt[3]{-\frac{N_2}{2} + \sqrt{\left(\frac{N_2}{2}\right)^2 + \left(\frac{N_1}{3}\right)^3}} + \sqrt[3]{-\frac{N_2}{2} - \sqrt{\left(\frac{N_2}{2}\right)^2 + \left(\frac{N_1}{3}\right)^3}} - \frac{\alpha_2}{3\alpha_1} \tag{4.11}$$

where

$$N_1 = \frac{a_3}{a_1} - \frac{1}{3}\left(\frac{a_2}{a_1}\right)^2, \; N_2 = \frac{2}{27}\left(\frac{a_2}{a_1}\right)^3 - \frac{1}{3}\left(\frac{a_2}{a_1}\right)\left(\frac{a_3}{a_1}\right) + \frac{a_4}{a_1}.$$

Hence, the corrosion rate can be expressed as:

$$\rho = \frac{x^2 - 1}{n - 1} \tag{4.12}$$

Therefore, the radial loss of steel at this stage can be expressed as:

$$\delta_{\text{stress}} = (d - d_\rho)/2 \tag{4.13}$$

where $d_\rho = \sqrt{1-\rho}\cdot d$ is the residual diameter of the steel bar after corrosion.

Let $q = q^{\text{inner}}$; the radial loss of steel bar $\delta_{\text{stress}}^{\text{inner}}$ at the inner cracking can be obtained using Eqs. (4.7)–(4.13).

4.4 PARTIAL CRACKING STAGE OF CORROSION-INDUCED CONCRETE CRACKING PROCESS

After the initiation of cracks at the interface between the steel and the concrete, the cracks in the concrete cylinder propagate along the radial direction. Thus, the thick-walled cylinder can be divided into two coaxial cylinders. One is the inner cracked cylinder and the other is the outer intact cylinder, as shown in Fig. 4.1b, where R_c is the radius at the interface between the cracked and intact cylinders. The intact part ($R_c \le r \le b$) is mainly based on elastic mechanics. Damage mechanics are applied to deal with the cracked part ($R \le r \le R_c$) and a damage variable is used to represent the severity of damage at the different radius of the concrete cylinder.

4.4.1 Intact Part

For the outer intact concrete cylinder, the theory of elasticity is applied. The stresses and strains at any point of the intact concrete cylinder are expressed as follows:

$$\begin{cases} \sigma_r = \dfrac{q_{R_c} R_c^2}{b^2 - R_c^2}\left(1 - \dfrac{b^2}{r^2}\right) \\[2ex] \sigma_\theta = \dfrac{q_{R_c} R_c^2}{b^2 - R_c^2}\left(1 + \dfrac{b^2}{r^2}\right) \end{cases} \tag{4.14}$$

$$\begin{cases} \varepsilon_r = \dfrac{1+\nu_c}{E_c'} \dfrac{q_{R_c} R_c^2}{b^2 - R_c^2} \left(1 - 2\nu_c - \dfrac{b^2}{r^2}\right) \\ \varepsilon_\theta = \dfrac{1+\nu_c}{E_c'} \dfrac{q_{R_c} R_c^2}{b^2 - R_c^2} \left(1 - 2\nu_c + \dfrac{b^2}{r^2}\right) \end{cases} \tag{4.15}$$

where q_{R_c} is the radial pressure at the interface between the intact and cracked cylinders.

4.4.2 Cracked Part

For the inner cracked concrete cylinder, the damage of the concrete cover varies along the radial direction. The concrete damage near the steel bar is more severe than that of the concrete far from the steel bar. A damage variable based on Mohr–Coulomb failure criterion [12] and the Mazar damage model [13] is adopted here to describe the different damage along the radial direction of the concrete cover.

1. Stress and radial expansive pressure at the steel/concrete interface

   The governing equation for equilibrium in the axisymmetric case can be expressed as follows [14]:

$$\frac{d\sigma_r}{dr} + \frac{\sigma_r - \sigma_\theta}{r} = 0 \tag{4.16}$$

   According to Mohr–Coulomb failure criterion with a damage variable D, the following relationship between two stress components can be obtained [17]:

$$\sigma_\theta = \sigma_r \frac{1-(1-D)\sin\varphi}{1+(1-D)\sin\varphi} + \frac{2c\cos\varphi}{1+(1-D)\sin\varphi} \tag{4.17}$$

   where c and φ are the cohesive strength and internal friction angle of concrete, respectively. The values of c and φ can be obtained from the literature [18]:

$$\begin{cases} c = \dfrac{1}{2}\sqrt{f_t \cdot f_c} \\ \varphi = 90^\circ - \dfrac{360^\circ}{\pi} \cdot \arctan\sqrt{\dfrac{f_t}{f_c}} \end{cases} \tag{4.18}$$

where f_c is the compressive strength of the concrete. The damage variable D can be obtained elsewhere [17]:

$$D = 1 - \frac{\varepsilon_t(1 - A_t)}{\varepsilon_\theta} - \frac{A_t}{\exp[B_t(\varepsilon_\theta - \varepsilon_t)]} \tag{4.19}$$

where A_t and B_t are coefficients of the Mazar damage model, $0.7 < A_t < 1$, $10^4 < B_t < 10^5$.

Substituting Eq. (4.17) into Eq. (4.16) yields:

$$\sigma_r = \exp\left(-\int_R^r \frac{2(1-D)\sin\varphi}{[1+(1-D)\sin\varphi]x}\mathrm{d}x\right) \cdot \left(\int_R^r \frac{2c\cos\varphi}{\xi[1+(1-D)\sin\varphi]} \cdot \exp\left(\int_R^\xi \frac{2(1-D)\sin\varphi}{[1+(1-D)\sin\varphi]x}\mathrm{d}x \cdot\right)\mathrm{d}\xi + C_0\right) \tag{4.20}$$

Using the following boundary conditions:

$$\sigma_r = -q \quad \text{at} \quad r = R \tag{4.21}$$

$$\begin{cases} D = 0 \\ \sigma_r = -q_{R_c} \quad \text{at} \quad r = R_c \\ \varepsilon_\theta = \varepsilon_t \end{cases} \tag{4.22}$$

The integration constant C_0 in Eq. (4.20) could be calculated, and $C_0 = -q$.

Substituting Eq. (4.22) into Eq. (4.15) yields:

$$q_{R_c} = \frac{E'_c}{1+\nu_c} \cdot \frac{\varepsilon_t(b^2 - R_c^2)}{R_c^2} \cdot \frac{1}{1 - 2\nu_c + b^2/R_c^2} \tag{4.23}$$

Substituting Eq. (4.22) and C_0 into Eq. (4.20) yields:

$$q = q_{R_c}\left(\frac{R_c}{R}\right)^{\frac{2\sin\phi}{1+\sin\phi}} + \frac{c\cos\phi}{\sin\phi}\left[\left(\frac{R_c}{R}\right)^{\frac{2\sin\phi}{1+\sin\phi}} - 1\right] \tag{4.24}$$

Let $R_c = R$ and note that the expansive pressure at the inner cracking obtained by Eqs. (4.23) and (4.24) is equal to that obtained by Eq. (4.6). Therefore, the calculated results obtained during the noncracking stage and partial cracking stage are continuous.

2. Strain and displacement

For the inner cracked concrete cylinder, the cracks are assumed to be smeared and uniformly distributed on the circumference in the cracked cylinder [10]; therefore, the concrete in that region behaves like an orthotropic material. Because all cracks are assumed to be in the radial direction and to propagate outward radially, the elastic modulus of the concrete in the radial direction $E_{c,r}$ remains unchanged while that in

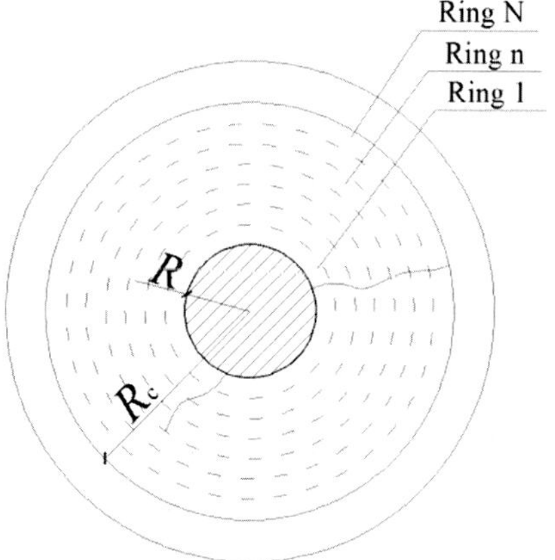

FIGURE 4.2 Partitions of the cracking part.

the hoop direction $E_{c,\theta}$ varies with the radius of the cylinder (ie, $E_{c,\theta} = (1 - D)E_{c,r}$, where the damage variable D is a function of the radial coordinate r).

To simplify the calculation, the cracked part is divided into N rings of equal thickness, as shown in Fig. 4.2, and the damage variable D within each ring is assumed to be constant.

The radial displacement u_r in the cracked concrete cylinder should satisfy the following equation [9]:

$$\frac{d^2 u_r}{dr^2} + \frac{1}{r}\frac{du_r}{dr} - k\frac{u_r}{r^2} = 0, \quad k = 1 - D \tag{4.25}$$

The solution to Eq. (4.25) can be expressed as:

$$u_r = C_1 r^{\sqrt{1-D}} + C_2 r^{-\sqrt{1-D}} \tag{4.26}$$

where C_1 and C_2 are constants. The radial and hoop strains can be calculated using Eq. (4.26) and are expressed as follows:

$$\begin{cases} \varepsilon_r = C_1\sqrt{1-D}r^{\sqrt{1-D}-1} - C_2\sqrt{1-D}r^{-\sqrt{1-D}-1} \\ \varepsilon_\theta = C_1 r^{\sqrt{1-D}-1} + C_2 r^{-\sqrt{1-D}-1} \end{cases} \tag{4.27}$$

Assuming that the thickness of each ring is ΔR, the damage process of the concrete cover can be described as follows: when the tensile strain in the hoop direction at the steel–concrete interface has reached the ultimate tensile strain of the concrete, that is, $\varepsilon_\theta|_{r=R} = \varepsilon_t$, the crack initiates at the steel–concrete interface and the crack length at the axial direction is ΔR. Hence, the thickness of the cracked part is ΔR. As the corrosion develops, the stress as well as the strain in concrete increase. When $\varepsilon_\theta|_{r=R+\Delta R} = \varepsilon_t$, the next ring cracks; the thickness of the cracked part is $2\Delta R$. The cracked part develops in this way until the cracks reach the surface of the concrete, that is, the thickness of the cracked part is equal to that of the concrete cover. This process can be computed as follows:

① Divide the cracked part. Assuming that the thickness of the cracked part is $N\Delta R$, $R_c = R + N\Delta R$. The cracked part is divided into N rings named Ring 1, Ring 2, ... Ring N − 1 and Ring N, respectively, as shown in Fig. 4.2.

② q and q_{R_c} can be obtained from Eqs. (4.23) and (4.24), respectively.

③ Ring N. Initiate from Ring N, and then calculate the rings inward in sequence.

Because Ring N is adjacent to the intact concrete zone, the damage variable of Ring N is very close to 0. Therefore, the damage variable D of the whole Ring N is assumed to be 0. Eq. (4.26) can be written as:

$$u_r^N = C_{2N-1}r + C_{2N}r^{-1} \tag{4.28}$$

By using Eqs. (4.15) and (4.27), the strains at $r = R_c$ can be expressed as follows:

$$\begin{cases} \varepsilon_r^N|_{r=R_c} = C_{2N-1} - C_{2N}r^{-2} = \dfrac{1+\nu_c}{E_c'}\dfrac{q_{R_c}R_c^2}{b^2-R_c^2}\left(1-2\nu_c-\dfrac{b^2}{R_c^2}\right) \\ \varepsilon_\theta^N|_{r=R_c} = C_{2N-1} + C_{2N}r^{-2} = \dfrac{1+\nu_c}{E_c'}\dfrac{q_{R_c}R_c^2}{b^2-R_c^2}\left(1-2\nu_c+\dfrac{b^2}{R_c^2}\right) \end{cases} \tag{4.29}$$

where C_{2N-1} and C_{2N} are constants, which can be obtained by solving Eq. (4.27), that is:

$$C_{2N-1} = \frac{1+\nu_c}{E_c'}\frac{q_{R_c}R_c^2}{b^2-R_c^2}(1-2\nu_c), \quad C_{2N} = \frac{1+\nu_c}{E_c'}\frac{q_{R_c}R_c^2}{b^2-R_c^2}b^2 \tag{4.30}$$

④ Ring N − 1.

By substituting Eq. (4.30) back into Eq. (4.28), the strain at the interface of Ring N and Ring N − 1 can be determined as follows:

$$\begin{aligned} \varepsilon_\theta^N|_{r=R_c-\Delta R} &= C_{2N-1} + C_{2N}(R_c-\Delta R)^{-2}, \\ \varepsilon_r^N|_{r=R_c-\Delta R} &= C_{2N-1} - C_{2N}(R_c-\Delta R)^{-2} \end{aligned} \tag{4.31}$$

The damage variable can be determined using the Mazar damage model [13]:

$$D|_{r=R_c-\Delta R} = 1 - \frac{\varepsilon_t(1-A_t)}{\varepsilon_\theta^N|_{r=R_c-\Delta R}} - \frac{A_t}{\exp[B_t(\varepsilon_\theta^N|_{r=R_c-\Delta R} - \varepsilon_t)]} \tag{4.32}$$

Like Ring N, let the damage variable of the whole Ring N − 1 equal that of the interface of Ring N and Ring N − 1, that is, $D^{N-1} = D|_{r=R_c-\Delta R}$. Thus, Eq. (4.26) can be written as:

$$u_r^{N-1} = C_{2N-3} r^{\sqrt{1-D^{N-1}}} + C_{2(N-1)} r^{-\sqrt{1-D^{N-1}}} \tag{4.33}$$

Similarly, the strains at $r = R_c - \Delta R$ can be expressed as follows:

$$\begin{cases} \varepsilon_r^{N-1}|_{r=R_c-\Delta R} = C_{2N-3}\sqrt{1-D^{N-1}}(R_c-\Delta R)^{\sqrt{1-D^{N-1}}-1} \\ -C_{2(N-1)}\sqrt{1-D^{N-1}}(R_c-\Delta R)^{-\sqrt{1-D^{N-1}}-1} = \varepsilon_r^N|_{r=R_c-\Delta R} \\ \varepsilon_\theta^{N-1}|_{r=R_c-\Delta R} = C_{2N-3}(R_c-\Delta R)^{\sqrt{1-D^{N-1}}-1} \\ +C_{2(N-1)}(R_c-\Delta R)^{-\sqrt{1-D^{N-1}}-1} = \varepsilon_\theta^N|_{r=R_c-\Delta R} \end{cases} \tag{4.34}$$

where C_{2N-3} and $C_{2(N-1)}$ are constants that can be obtained by solving Eq. (4.34):

$$\begin{aligned} C_{2N-3} &= \frac{\varepsilon_r^N|_{r=R_c-\Delta R} + \sqrt{1-D^{N-1}}\varepsilon_\theta^N|_{r=R_c-\Delta R}}{2\sqrt{1-D^{N-1}}(R_c-\Delta R)^{\sqrt{1-D^{N-1}}-1}}, \\ C_{2(N-1)} &= \frac{\sqrt{1-D^{N-1}}\varepsilon_\theta^N|_{r=R_c-\Delta R} - \varepsilon_r^N|r = R_c - \Delta R}{2\sqrt{1-D^{N-1}}(R_c-\Delta R)^{-\sqrt{1-D^{N-1}}-1}} \end{aligned} \tag{4.35}$$

⑤ Ring n.

Obviously, the radial displacement of Ring n + 1 has been obtained:

$$u_r^{n+1} = C_{2n+1} r^{\sqrt{1-D^{n+1}}} + C_{2(n+1)} r^{-\sqrt{1-D^{n+1}}} \tag{4.36}$$

The strains at the interface of Ring n and Ring n + 1 can be calculated as:

$$\begin{cases} \varepsilon_r^{n+1}|_{r=R+n\Delta R} = C_{2n+1}\sqrt{1-D^{n+1}}(R+n\Delta R)^{\sqrt{1-D^{n+1}}-1} \\ \qquad -C_{2(n+1)}\sqrt{1-D^{n+1}}(R+n\Delta R)^{-\sqrt{1-D^{n+1}}-1} \\ \varepsilon_\theta^{n+1}|_{r=R+n\Delta R} = C_{2n+1}(R+n\Delta R)^{\sqrt{1-D^{n+1}}-1} \\ \qquad +C_{2(n+1)}(R+n\Delta R)^{-\sqrt{1-D^{n+1}}-1} \end{cases} \tag{4.37}$$

The damage variable can be expressed as:

$$D^n = D|_{r=R+n\Delta R} = 1 - \frac{\varepsilon_t(1-A_t)}{\varepsilon_\theta^{n+1}|_{r=R+n\Delta R}} - \frac{A_t}{\exp[B_t(\varepsilon_\theta^{n+1}|_{r=R+n\Delta R} - \varepsilon_t)]} \tag{4.38}$$

For Ring n, Eq. (4.26) can be written as:

$$u_r^n = C_{2n-1} r^{\sqrt{1-D^n}} + C_{2n} r^{-\sqrt{1-D^n}} \tag{4.39}$$

The strains at $r = R_c + n\Delta R$ can be expressed as follows:

$$\begin{cases} \varepsilon_r^n|_{r=R+n\Delta R} = C_{2n-1}\sqrt{1-D^n}(R+n\Delta R)^{\sqrt{1-D^n}-1} \\ -C_{2n}\sqrt{1-D^n}(R+n\Delta R)^{-\sqrt{1-D^n}-1} = \varepsilon_r^{n+1}|_{r=R+n\Delta R} \\ \varepsilon_\theta^n|_{r=R+n\Delta R} = C_{2n-1}(R+n\Delta R)^{\sqrt{1-D^n}-1} \\ +C_{2n}(R+n\Delta R)^{-\sqrt{1-D^n}-1} = \varepsilon_\theta^{n+1}|_{r=R+n\Delta R} \end{cases} \tag{4.40}$$

The constants C_{2n-1} and C_{2n} can be obtained by solving Eq. (4.40):

$$\begin{aligned} C_{2n-1} &= \frac{\varepsilon_r^{n+1}|_{r=R+n\Delta R} + \sqrt{1-D^n}\varepsilon_\theta^{n+1}|_{r=R+n\Delta R}}{2\sqrt{1-D^n}(R+n\Delta R)^{\sqrt{1-D^n}-1}}, \\ C_{2n} &= \frac{\sqrt{1-D^n}\varepsilon_\theta^{n+1}|_{r=R+n\Delta R} - \varepsilon_r^{n+1}|_{r=R+n\Delta R}}{2\sqrt{1-D^n}(R+n\Delta R)^{-\sqrt{1-D^n}-1}} \end{aligned} \tag{4.41}$$

⑥ Ring 1.

By repeating step ⑤ from Ring N to Ring 1 in sequence, the radial displacement of Ring 1 can finally be obtained as:

$$u_r^1 = C_1 r^{\sqrt{1-D^1}} + C_2 r^{-\sqrt{1-D^1}} \tag{4.42}$$

3. Steel corrosion

The deformation of concrete at the steel–concrete interface is equal to u_r^1:

$$\delta_c = u_r^1 = C_1 R^{\sqrt{1-D^1}} + C_2 R^{-\sqrt{1-D^1}} \tag{4.43}$$

The radial loss of steel at this stage can also be calculated using Eqs. (4.7)–(4.13).

When the cracks reach the surface of the concrete cylinder (ie, $R_c = b$), the radial loss of steel calculated with the aforementioned method is thus the steel bar radial loss $\delta_{stress}^{surface}$.

4.5 CORROSION-INDUCED EXPANSIVE PRESSURE

The corrosion-induced expansive pressure between the rust layer and the concrete directly causes the concrete cover to crack. However, it has rarely been discussed previously. The expansive pressure induced by steel corrosion is studied here based on the aforementioned model. The factors affecting the

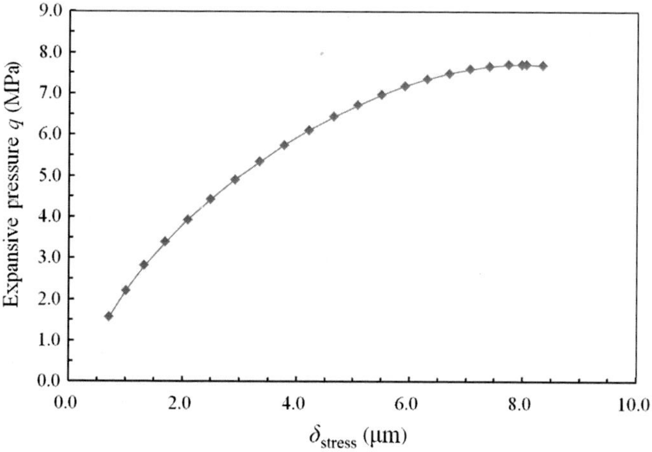

FIGURE 4.3 Expansive pressure against steel corrosion.

expansive pressure are discussed, including concrete cover thickness, steel bar diameter, and tensile strength of the concrete.

4.5.1 Relation Between Expansive Pressure and Steel Corrosion

After steel depassivation, the concrete cover is in the noncracking stage, as shown in Fig. 4.1a. During this stage, the expansive pressure can be calculated by Eq. (4.7). Once the cracks at the concrete inner interface occur, the concrete cover moves into the partial cracking stage, as illustrated in Fig. 4.1b. Considering different damage along the radial direction of the concrete cover, the expansive pressure has been derived as Eqs. (4.23) and (4.24). The assumption and analysis methods for each stage are different. However, the expansion pressure of two stages is consistent, as stated in Section 4.4.2.

To show the relationship between the expansive pressure and the steel corrosion after the inner cracking of the concrete cover, the reinforced concrete cylinder is used as an example assuming $f_c = 35$ MPa, $C = 30$ mm, $d = 16$ mm, $n = 2.6$, $v_c = 0.3$, $E_r = 100$ MPa, $v_r = 0.25$, $A_t = 0.7$, $B_t = 10,000$, and $\Delta R = 0.3$ mm. Other variables could be calculated according to Chinese Code [19] and Eq. (4.18) as $f_t = 2.18$ MPa, $E_c = 31.3$ GPa, $E'_c = 15.7$ GPa, $c = 4.37$ MPa, and $\varphi = 62.0°$. Fig. 4.3 illustrates that as corrosion products accumulate, the expansive pressure increases continuously, from the noncracking stage until the partial cracking stage.

4.5.2 Variation of Expansive Pressure

Using the same example stated in Section 4.5.1, the expansive pressure as a function of crack length during the partial cracking stage is studied.

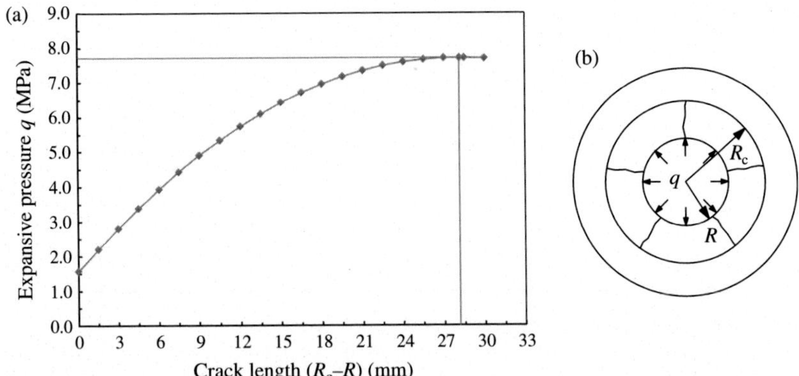

FIGURE 4.4 Variation of expansive pressure after initiation of cracks in concrete cover. (a) Expansive pressure against crack length. (b) Expansive pressure in cracked concrete.

Fig. 4.4 illustrates the variation of the expansive pressure after the initiation of cracks in the concrete cover. It can be seen that the expansive pressure increases with the propagation of the cracks; however, this increasing trend ends before the cracks reach the surface. After the expansive pressure gets to the peak value, it decreases slowly with the development of the cracks. This phenomenon was mentioned previously by Li [9]. It should be noted that after the expansive pressure reaches the maximum value, the whole concrete cover will crack spontaneously, even without more steel corrosion growth. This is because as the steel corrosion grows, the elastic strain energy accumulates in the concrete cover. When the elastic strain energy of the concrete cover is greater than the energy required to crack the concrete, the cracks appear at the inner surface of the concrete cover. After the initiation of cracks, with the growth of the corrosion products, the strain energy of the concrete cover develops, leading to the propagation of the cracks. Because the concrete is a brittle material, once the strain energy gets to a certain extent, the concrete cover is unable to absorb more energy induced by the expansive pressure, thus resulting in prompt propagation of cracks to the concrete surface. Therefore, the moment when the expansive pressure gets to the peak value is considered to correspond to concrete surface cracking. After this moment, the path for load transmission reduces due to the existence of the cracks that pass through the whole concrete cover, inducing the decline of the expansive pressure.

4.5.3 Effect of Concrete Cover Thickness

To examine the effect of the concrete cover thickness on the expansive pressure, four reinforced concrete examples are calculated here. The thicknesses of the concrete covers are 20, 25, 30, and 35 mm, respectively, whereas all other parameters are kept the same as those used in Section 4.5.1.

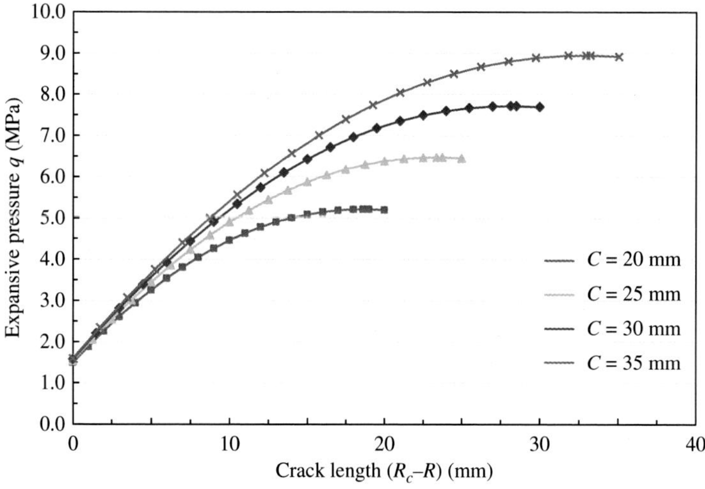

FIGURE 4.5 Effect of concrete cover thickness on expansive pressure.

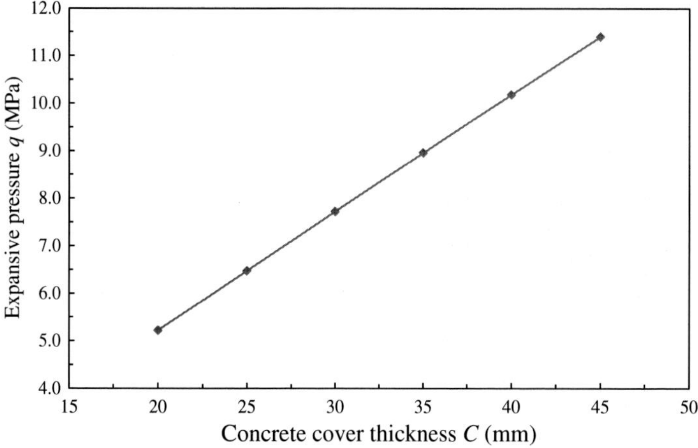

FIGURE 4.6 Peak value of expansive pressure against concrete cover thickness.

The expansive pressures as a function of crack length for different cover thickness are shown in Fig. 4.5, which indicates that the thicker the concrete cover, the larger the peak value of the expansive pressure. Fig. 4.6 depicts the maximum value of expansive pressure $q_{\max}$ as a linear function of concrete cover thickness. This is understandable because the thicker the concrete cover, the more strain energy that is needed to crack the concrete, thus resulting in larger expansive pressure.

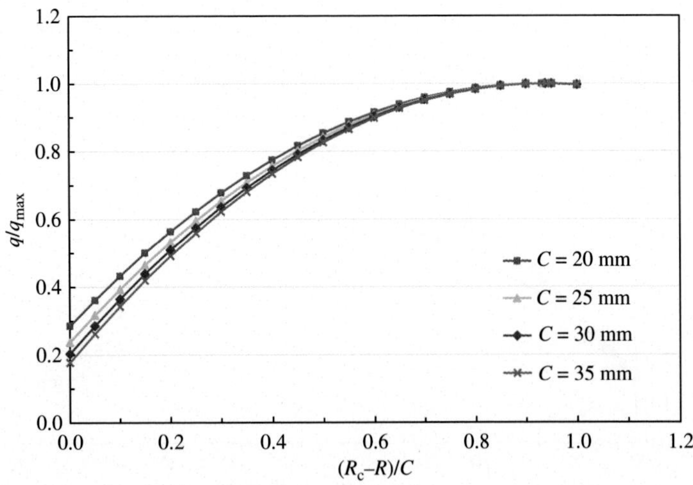

FIGURE 4.7 Normalized expansive pressure as a function of normalized crack length.

Fig. 4.7 presents the normalized expansive pressure versus the normalized crack length. In this figure, the vertical coordinate is q/q_{max} (the ratio of the expansive pressure and the maximum expansive pressure) and is called normalized expansive pressure. The horizontal coordinate is $(R_c - R)/C$ (the ratio of the crack length and the concrete cover thickness) and is called the normalized crack length. It can be seen that the relations for different cover thicknesses do not show much difference, especially when the cracks propagate near the surface of the concrete cover. The peak values of all four examples occur at a position of approximately $0.94C$. Based on these results, if the other parameters are fixed, then the position of the maximum expansive pressure relative to the thickness of concrete cover is nearly the same.

4.5.4 Effect of Steel Bar Diameter

To examine the effect of the steel bar diameter on the expansive pressure, we used the same reinforced concrete example as that used in Section 4.5.1. Steel bar diameters were 12, 16, 20, and 25 mm, whereas all other parameters were kept unchanged.

The expansive pressures as a function of crack length for different steel bar diameters are shown in Fig. 4.8. It can be seen that the expansive pressure decreases as the steel bar diameter increases. This is because for the same concrete cover thickness, the same steel loss of the larger steel bar diameter produces larger strain energy in the concrete. Therefore, the steel bar with the larger diameter is able to crack the concrete cover easily compared with the bar with the smaller diameter.

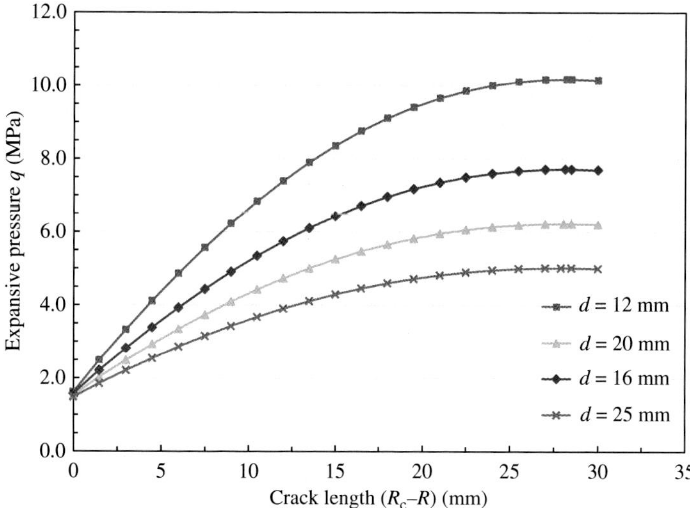

FIGURE 4.8 Effect of steel bar diameter on expansive pressure.

However, although the steel bar diameter affects the value of the expansive pressure, it has little influence on the position of the maximum expansive pressure, as shown in Fig. 4.8.

4.5.5 Effect of Concrete Quality

To examine the effect of concrete quality on the expansive pressure, the same reinforced concrete example in Section 4.5.1 is used. Normally, the tensile strength is considered to be a key factor in corrosion-induced cracking of concrete. Concrete with a larger tensile strength always has a larger compressive strength. We chose four compressive strengths rather than tensile strengths to illustrate the effect of concrete quality on expansive pressure (ie, 25, 35, 45, and 55 MPa); all other parameters are the same as those mentioned in Section 4.5.1. According to Chinese code [19] and Eq. (4.18), the elastic modulus, E_c, deformation modulus at the maximum tensile stress, E_c', tensile strength, f_t, cohesive strength, c, and the internal friction angle, φ, of the concrete could be calculated as shown in Table 4.1.

Fig. 4.9 shows the effect of the concrete quality on the expansive pressure. It can be seen from the figure that the peak value of the expansive pressure increases with the strength of concrete. That is because that better quality of concrete, illustrated by higher tensile strength, will heighten the bearing capacity of the concrete cover.

TABLE 4.1 Calculated Mechanical Parameters of the Four Types of Concrete

f_c (MPa)	E_c (GPa)	E'_c (GPa)	f_t (MPa)	c (MPa)	φ (°)
25	27.9	13.9	1.81	3.37	59.8
35	31.3	15.7	2.18	4.37	62.0
45	33.7	16.8	2.51	5.31	63.4
55	35.3	17.7	2.80	6.20	64.6

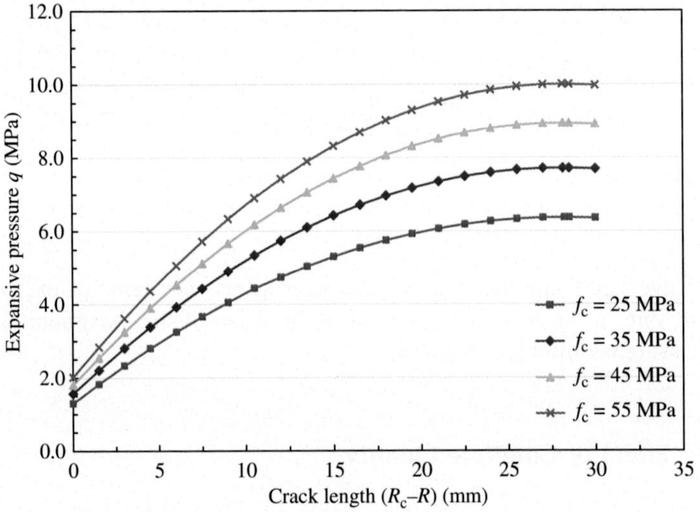

FIGURE 4.9 Effect of tensile strength on expansive pressure.

4.6 DISCUSSION ON THE RADIAL LOSS OF STEEL BAR

Compared with the expansion pressure, the radial loss of steel bar, which is easy to be tested, is given more attention by researchers and engineers. The quantitative relationship between the radial loss of steel bar and concrete cover cracking has been widely discussed in the literature [20–24]. Here, the radial loss of steel bar is also discussed based on the model proposed.

4.6.1 Steel Loss Varying with the Crack Length

The steel loss at any time during the corrosion-induced cracking process can be obtained using the proposed model. The same reinforced concrete example used in Section 4.5.1 is used here to study the steel loss during the concrete cracking process. Fig. 4.1 shows that steel loss has a nearly linear relationship with the crack length in the concrete cover.

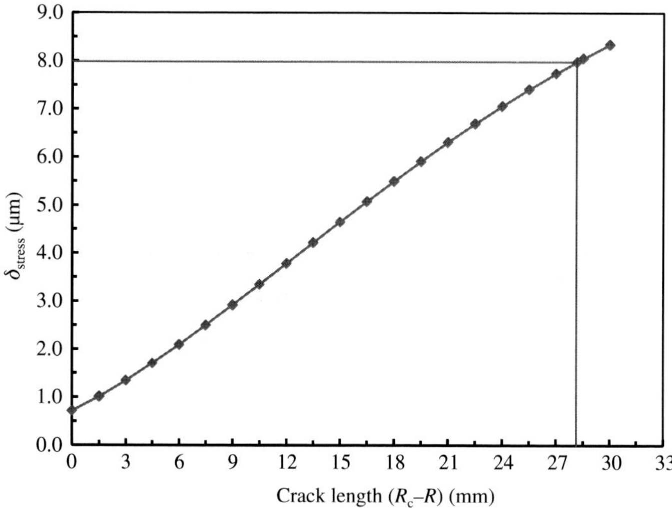

FIGURE 4.10 Radial loss of steel bar as a function of crack length.

According to the discussion in Section 4.5.2, the steel loss, δ_{stress}, during concrete surface cracking should correspond to the moment when the expansive pressure gets to the peak value, but not when $R_c - R = 30$ mm. In this case, it is when $R_c - R = 0.94C = 28.1$ mm, as shown in Fig. 4.10. The factors affecting the steel loss during concrete surface cracking are discussed using the same example stated in Section 4.5.1.

4.6.2 Effect of Concrete Cover Thickness

Let concrete cover thickness, C, equal 20, 25, 30, 35, 40, and 45 mm while all other parameters are kept unchanged. The effect of concrete cover thickness on the steel loss during surface cracking is shown in Fig. 4.11. It is clear that with the concrete cover thickness increases, the steel loss at surface cracking increases. According to the discussion in Section 4.5.3, the thicker the concrete cover, the larger the maximum expansive pressure. Therefore, the steel loss during surface cracking will be larger.

4.6.3 Effect of Steel Bar Diameter

Use steel bar diameters of 12, 16, 20, and 25 mm while keeping all other parameters unchanged. The influence of the steel bar diameter on the steel loss during surface cracking is shown in Fig. 4.12. With the increase of steel bar diameter, the steel loss during surface cracking decreases. Therefore, the increase of the steel bar diameter will result in the earlier cracking of the concrete surface.

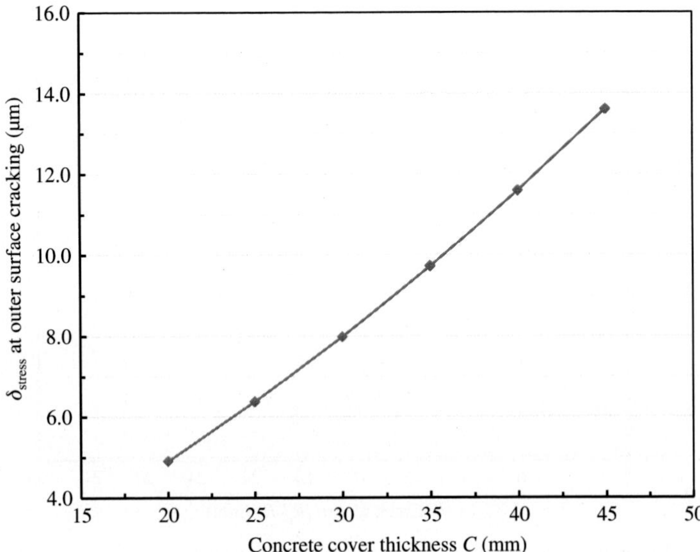

FIGURE 4.11 Effect of concrete cover thickness on steel loss during surface cracking.

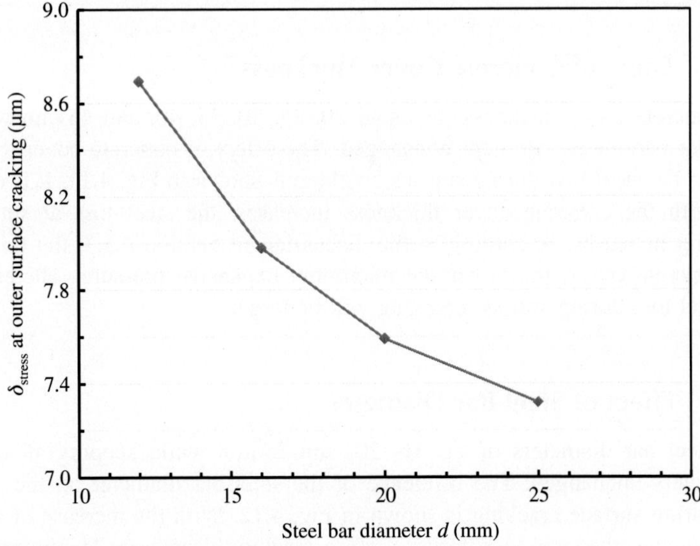

FIGURE 4.12 Effect of steel bar diameter on steel loss during surface cracking.

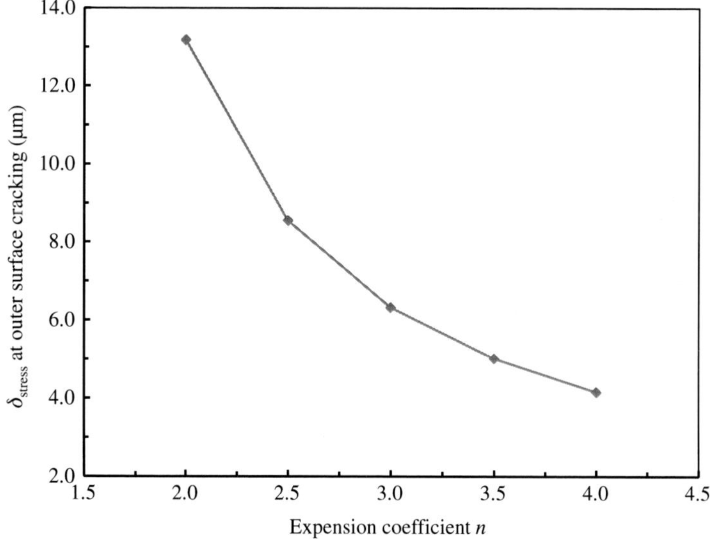

FIGURE 4.13 Effect of rust expansion coefficient on steel loss during surface cracking.

4.6.4 Effect of Rust Expansion Coefficient

Fig. 4.13 indicates the rust expansion coefficient significantly effects on the steel loss during surface cracking. When the rust expansion coefficient increases, the steel loss during surface cracking decreases quickly. Therefore, to reasonably predict the steel loss at concrete surface cracking, obtaining the real rust expansion coefficient (not assumed value) is very important. The rust expansion coefficient n is mainly influenced by the environment of reinforced concrete structures (see Section 3.2.4.4 "Rust expansion coefficients in different environments" in chapter "The Expansion Coefficients and Modulus of Steel Corrosion").

4.6.5 Effect of Concrete Quality

Parameters in Section 4.5.5 are used to examine the effect of concrete quality on the radial loss of steel bar at the concrete outer surface cracking moment; the results are shown in Fig. 4.14. Steel radial loss at the surface cracking moment increases with the growth of concrete strength. Therefore, the decrease of concrete strength will result in earlier cracking of the concrete surface.

4.7 CONCLUSIONS

An analytical model for the corrosion-induced concrete cracking process in reinforced concrete structures has been developed in this chapter. The stresses and cracks in the concrete cover that developed due to steel corrosion have

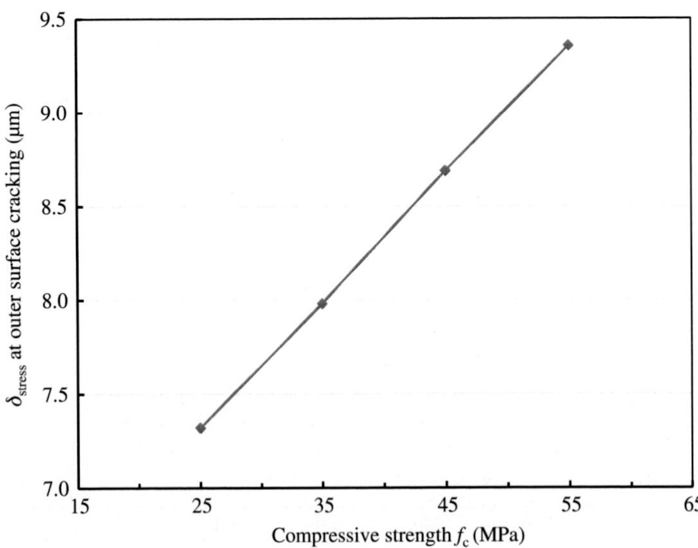

FIGURE 4.14 Effect of compressive strength on steel loss during surface cracking.

been studied. Elastic mechanics were utilized to analyze the whole concrete cover subjected to corrosion expansion during the early stage. Once the inner crack between the concrete and steel occurred, the concrete cover was divided into two regions, the inner cracked part and the outer intact part. Elastic mechanics were used for analyzing the intact part and damage mechanics were applied to deal with the cracked part. The damage variable, based on Mohr–Coulomb failure criterion and the Mazar damage model, was adopted to describe the various damage along the radial direction of the concrete cover.

Based on this model, the expansive pressure induced by steel corrosion and steel loss during concrete surface cracking were discussed. It was found that as corrosion products accumulate, the expansive pressure increases, at first. However, during the partial cracking stage, after the expansive pressure got to the peak value, it decreased slowly with the development of the cracks. Before surface cracking, as cracks propagate, the steel loss increased almost linearly.

The parameters affecting the expansive pressure and the steel loss during concrete surface cracking, including concrete cover thickness and quality, steel bar diameters, and rust expansion coefficients, were also discussed.

REFERENCES

[1] Bazant ZP. Physical model for steel corrosion in concrete sea structures–application. J Struct Div 1979;105 [ASCE 14652 Proceeding]

[2] Liu Y, Weyers RE. Modeling the time-to-corrosion cracking in chloride contaminated reinforced concrete structures. ACI Mater J 1998;95(6):675–81.

[3] Zhao YX, Jin WL. Modeling the amount of steel corrosion at the cracking of concrete cover. Adv Struct Eng 2006;9(5):687–96.
[4] El Maaddawy T, Soudki K. A model for prediction of time from corrosion initiation to corrosion cracking. Cement Concrete Composites 2007;29(3):168–75.
[5] Malumbela G, Alexander M, Moyo P. Model for cover cracking of RC beams due to partial surface steel corrosion. Constr Building Mater 2011;25(2):987–91.
[6] Kim KH, Jang SY, Jang BS, et al. Modeling mechanical behavior of reinforced concrete due to corrosion of steel bar. ACI Mater J 2010;107(2):106–13.
[7] Lu C, Jin W, Liu R. Reinforcement corrosion-induced cover cracking and its time prediction for reinforced concrete structures. Corrosion Sci 2011;53(4):1337–47.
[8] Bhargava K, Ghosh AK, Mori Y, et al. Modeling of time to corrosion-induced cover cracking in reinforced concrete structures. Cement Concrete Res 2005;35(11):2203–18.
[9] Li C, Melchers RE, Zheng J. Analytical model for corrosion-induced crack width in reinforced concrete structures. ACI Struct J 2006;103(4):479–87.
[10] Pantazopoulou SJ, Papoulia KD. Modeling cover-cracking due to reinforcement corrosion in RC structures. J Eng Mech 2001;127(4):342–51.
[11] Chernin L, Val DV, Volokh KY. Analytical modelling of concrete cover cracking caused by corrosion of reinforcement. Mater Struct 2010;43(4):543–56.
[12] Mohr O. Welche Umstände bedingen die Elastizitätsgrenze und den Bruch eines Materials. Zeitschrift Ver Dtsch Ingenieure 1900;46:1524–30.
[13] Lemaitre J. A course on damage mechanics. 2nd ed. Berlin: Springer; 1990.
[14] Timoshenko S, Goodier JN. Theory of elasticity. New York, NY: Hill Book Co.; 1970.
[15] Guo ZH. The strength and deformation of concrete—experimental basis and constitutive relation. Beijing: Tsinghua University Press; 1999 [in Chinese].
[16] Shu SL. Reinforced concrete structures (Edition 2). Hangzhou: Zhejiang University Press; 2003 [in Chinese].
[17] Shen XP, Yang L. Concrete damage theory and experiments. Beijing: Science Press; 2009 [in Chinese].
[18] Li YA, Ge XR, Mi GR, Zhang HC. Failure criteria of rock-soil-concrete and estimation of their strength parameters. Chin J Rock Mech Eng 2004;23(5):770–6 [in Chinese].
[19] Ministry of Housing and Urban-Rural Development of the People's Republic of China (MOHURD). Code for design of concrete structures. Beijing: China China Architecture and Building Press; 2010 [in Chinese].
[20] Andrade C, Alonso C, Molina FJ. Cover cracking as a function of bar corrosion: Part I-experimental test. Mater Struct 1993;26(8):453–64.
[21] Alonso C, Andrade C, Rodriguez J, et al. Factors controlling cracking of concrete affected by reinforcement corrosion. Mater Struct 1998;31(7):435–41.
[22] Zhang WP. Damage prediction and durability estimation for corrosion of reinforcement in concrete structures. Shanghai: Tongji University; 1999 [in Chinese].
[23] Vidal T, Castel A, Francois R. Analyzing crack width to predict corrosion in reinforced concrete. Cement Concrete Res 2004;34(1):165–74.
[24] Zhang R, Castel A, François R. Concrete cover cracking with reinforcement corrosion of RC beam during chloride-induced corrosion process. Cement Concrete Res 2010;40 (3):415–25.

Chapter 5

Mill Scale and Corrosion Layer at Concrete Surface Cracking

Chapter Outline

5.1 INTRODUCTION

When steel corrosion grows in a concrete cover, cracks initiate at the steel–concrete interface (ie, inner cracking) and propagate outward to the surface of the concrete cover (ie, surface cracking). These cracks provide a path for the rapid ingress of aggressive agents to the reinforcement, which can accelerate the corrosion process.

Because concrete is a brittle material, once the strain energy gets to a certain extent, the concrete cover is unable to absorb more energy induced by the expansive pressure, resulting in prompt propagation of cracks to the concrete surface. The inner cracking time (ie, the moment when the inner surface of concrete cover cracks) may be close to the moment of outer surface cracking of concrete. Estimating the inner cracking time of the concrete cover is helpful for the prediction of concrete outer surface cracking. Therefore, we investigated the critical thickness of the corrosion layer at the inner as well as the outer concrete surface cracking in the chapter.

Previous studies [1,2] have shown that mill scale exists in corroded steel bars. Because the mill scale formed before corrosion initiation but was not generated during the rust expansion process [3], it does not actually contribute to the rust volume expansion acting on the surrounding concrete cover. Hence, the mill scale cannot be taken into account in the concrete cracking

Steel Corrosion-Induced Concrete Cracking. DOI: http://dx.doi.org/10.1016/B978-0-12-809197-5.00005-0

model induced by steel corrosion. Previous studies, however, rarely considered the effect of mill scale when developing the empirical models based on the tested data. The work in this chapter also investigates the influences of mill scale on the concrete cracking model.

5.2 EXPERIMENTAL PROGRAM

5.2.1 Reinforced Concrete Specimens

Three cylindrical reinforced concrete specimens of the dimensions shown in Fig. 5.1 were used in this study. Each specimen had a diameter of 70 mm and a height of 150 mm. Each specimen contained a plain steel bar of 16 mm in the center. The rebars were used as received and no efforts were made to remove the existing mill scale. The mixture proportions of the concrete are reported in Table 5.1.

The specimens were covered in damp hessian immediately after casting and sprayed with water once every day for 2 weeks. Thereafter, they were dried at room temperature (approximately 25°C) for several weeks. The 28-day compressive strength of the concrete was 19.4 MPa as measured on 150-mm cubes.

Both ends of each specimen and the outer steel bars were surface-treated with epoxy resin, as shown in Fig. 5.1, to avoid chloride ingress from the ends during the accurate corrosion process.

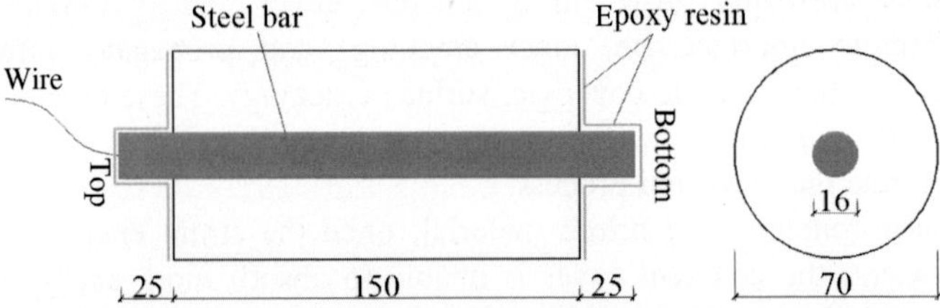

FIGURE 5.1 Layout details of the concrete specimens (dimensions are in mm).

TABLE 5.1 Mixture Composition of the Concrete Specimens (kg/m^3)

Ratio of Water/Binder	Cement	Aggregate	Sand	Fly Ash	Slag	Silica	Water-Reducing Admixtures	Water
0.55	207	1197	631	52	52	35	0.833	155

5.2.2 Accelerated Steel Corrosion

After immersion in a 3.5% NaCl solution for 24 h, the test specimens were wrapped in stainless steel nets. A constant current was applied between the reinforcing steel bars (acting as the anode) in the specimen and the outside stainless steel net (acting as the cathode) controlled by a DC power source. The currents applied were 300 $\mu A/cm^2$.

The evolution of cracks on the surface of the specimen was observed twice per day. On day 22, a corrosion-induced crack was observed on one side of specimen 1, running approximately parallel to the center reinforcing bar. The longitudinal length of the crack was 45 mm and the crack width was approximately 0.05 mm. On day 28.5, a longitudinal crack 37 mm in length and 0.05 mm in width was observed on one side of specimen 2, and a short crack 7 mm in length with a crack width of approximately 0.05 mm was observed on one side of specimen 3. The specimens were removed for further investigation at the moment mentioned.

5.2.3 Sample Preparation

5.2.3.1 Samples for Digital Microscopy Observation

The thickness of the rust layer was measured by a digital microscope. The sample preparation for digital microscopy observation is presented here.

The cracking parts of specimens were cast into a low-viscosity epoxy resin to minimize any artificial damage that might be incurred during the sample preparation process. As shown in Fig. 5.2, an adhesive tape was stuck around the cracked end of the specimen, and then the low-viscosity epoxy resin was poured into the enclosed area on the top end of the specimen. The epoxy was allowed to harden for several days; afterward, the adhesive tape was removed and cutting was performed to obtain a 10-mm-thick slice using an abrasive cutter and diamond blade suitable for hard brittle materials at a very low feed rate of less than 0.05 mm/s with water as a coolant to ensure minimal damage. After one slice was obtained, the same process was repeated to make

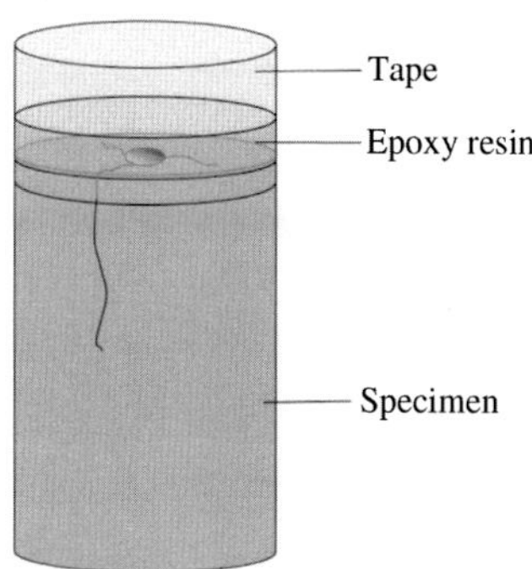

FIGURE 5.2 Cracking parts of specimens were cast into a low-viscosity epoxy resin.

other slices. That means this process (ie, tape sticking, epoxy pouring, tape removing, and cutting) was performed to make each slice.

The observed results show that cracking at the ends of the specimens is more severe, as illustrated in Fig. 5.3. Because the purpose of this experiment is to measure the thickness of the rust layer at concrete surface cracking, the severe corroded slices near the specimen ends are not of interest for this study. However, the slices near the tips of the longitudinal corrosion-induced cracks, which are selected for further investigation, are useful for achieving the aim of this study. The locations of the chosen slices for observation and measurements are also illustrated in Fig. 5.3. Each specimen was sectioned to produce a series of 10-mm-thick cross-section slices. Fig. 5.4 shows a sample (slice 1-1) for observation with a digital microscope.

The slices were all polished using a lapping and polishing machine. To prevent further corrosion, slices were kept in a dry environment (relative humidity less than 30%) for 1–5 days before observation.

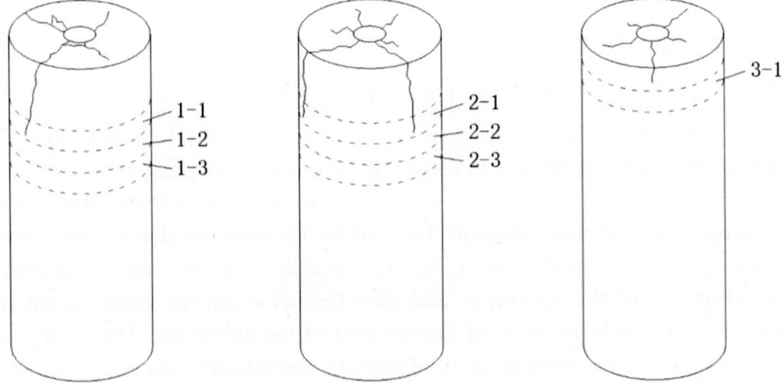

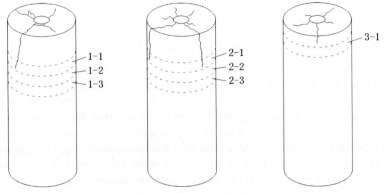

FIGURE 5.3 Schematic diagrams of the specimens and the location of the slices.

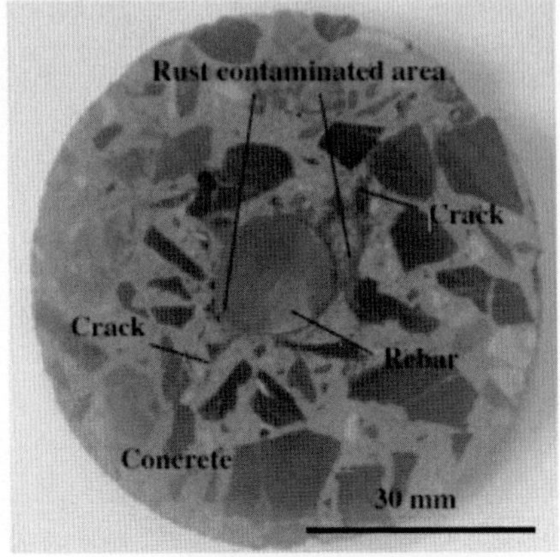

FIGURE 5.4 Sample (slice 1-1) for digital microscope observation.

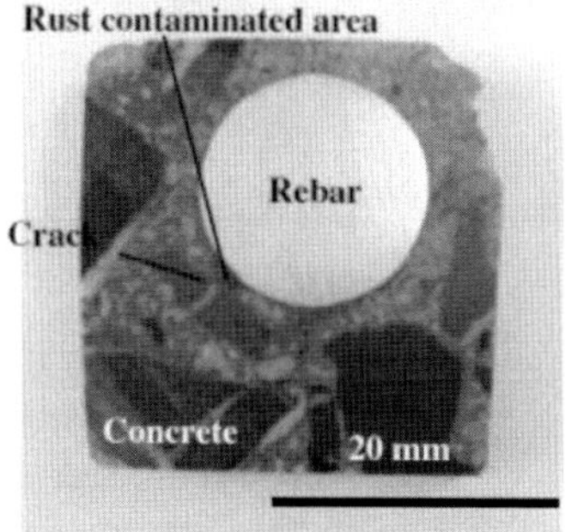

FIGURE 5.5 Sample trimmed from slice 2-2 for SEM observation.

5.2.3.2 Sample for SEM Observation

The mill scale and corrosion products generated after casting of the specimens cannot be distinguished by digital microscopy. Therefore, a sample was prepared for scanning electron microscope (SEM) observation. Slice 2-2 was trimmed to a smaller block including the steel bar and the cracks (Fig. 5.5) and then ground with SiC papers at 400, 1000, and 2000 grit to expose a fresh surface. The dimensions of this sample after being cut and ground were $25 \times 25 \times 8$ mm.

This sample was also used to observe the rust distribution in corrosion-induced cracks and the adjacent areas.

5.2.4 Observation and Measurement

5.2.4.1 Rust Layer

The boundaries of the rebar, rust layer, and concrete of each slice were observed by a digital microscope (Pro-micro scan 5866). The rust layer could be clearly distinguished in color images when the digital microscope connected with the computer, which allowed for accurate measurements of the corrosion layer thickness. The area of the rust layer was also measured by the digital microscope.

5.2.4.2 Steel–Concrete Interface

A SEM (Hitachi S3400N) operated in the backscattered electron (BSE) mode was used to study the steel–concrete interface. The iron oxides that accumulated at the steel–concrete interface were investigated, and the contents of Fe and O were analyzed by energy dispersive spectroscopy (EDS). The boundaries of the steel, corrosion layer (CL), mill scale (MS), corrosion products-filled paste (CP), and concrete could be distinguished by SEM images and EDS analysis.

5.2.4.3 *Rust Distributions in Corrosion-Induced Cracks*

A SEM (Hitachi S3400N) operated in the BSE mode was used to study the rust distributed in the corrosion-induced cracks. The EDS analysis was applied to study the content of Fe and O in the observation areas.

5.3 RUST DISTRIBUTIONS IN THE CRACKING SAMPLE

Fig. 5.6 illustrates the rust distribution at the steel–concrete interface and in the cracks that was caused by expansive stresses exerted by the corrosion products. It can be seen in Fig. 5.6a that some parts of the concrete adjacent to the corroding steel, which is defined as corrosion products-filled paste (CP) in this book, are brighter than the concrete farther away (seen more clearly in Fig. 5.7a). This suggests an increase in the local mean atomic number and backscattering coefficient due to ingress of Fe into the concrete. This phenomenon has been found by other researchers [1,2,4] and is discussed in detail in chapters "Rust Distribution in Corrosion-Induced Cracking Concrete" and "Development of the Corrosion Products-Filled Paste at the Steel/Concrete Interface."

Fig. 5.6b shows that although rust does not fill the top right crack, Fe enters the concrete at both sides of this crack (these areas are brighter). The bottom left crack remains empty. These are the only two kinds of rust distribution in cracks; no rust was found in the cracks in this sample. This is confirmed by the EDX analysis in Fig. 5.7.

The linear distribution of Fe across a corrosion-induced crack was analyzed by EDS along an analytical line, as shown in Fig. 5.7. In Fig. 5.7b, the horizontal axis represents the distance from the starting point of the analytical line, whereas the vertical axis indicates the counts per second of photoelectrons, reflecting the content of Fe. It can be seen that the rust penetrates the concrete, and high concentrations of Fe are shown in the contaminated area. However, a deep concave curve in one region of the analytical line, which corresponds with

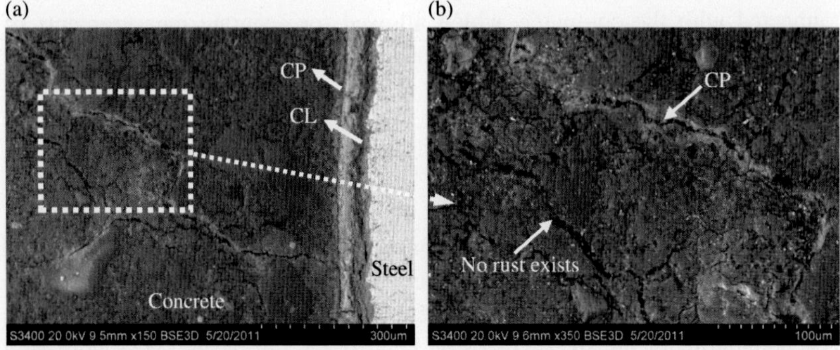

FIGURE 5.6 Rust distributions at the steel–concrete interface and in the corrosion-induced cracks (*CP*, corrosion products-filled paste; *CL*, corrosion layer).

(a)

(b)

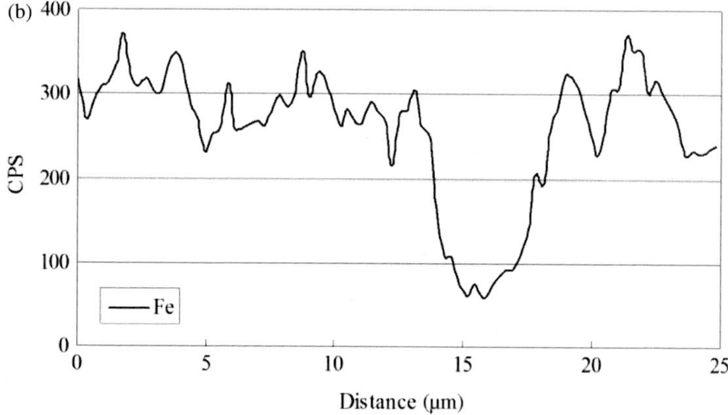

FIGURE 5.7 EDS analysis across a corrosion-induced crack. (a) Corrosion-induced crack and an analytical line across the crack. (b) Distribution of Fe across the corrosion-induced crack analyzed by EDS along the analytical line.

the location of the crack, indicates much less Fe in the crack. This result again illustrates that corrosion products do not fill the cracks.

Based on this observation from the sample, it can be proposed that the rust does not fill the cracks in the corrosion-induced cracking process. However, the rust was observed to penetrate the concrete at both sides of cracks. The reason for the existing rust in the contaminated concrete is that the outer solution ingresses into the cracks after concrete surface cracking during the acceleration corrosion process and some rust dissolves in the solution and is carried away from the rebar by the solution, lining and penetrating the edges of the cracks.

According to these observations, it can be supposed that the filling of corrosion-induced cracks with rust does not need to be considered before concrete outer surface cracking.

5.4 MILL SCALE

Mill scale is a thin, adherent oxide coating that forms on steel during heat treatment, hot rolling, and forging processes; the oxidation of the steel occurs during exposure to air and while cooling from the rolling temperature [3]. The Fe-to-O atomic ratio of mill scale is higher than that of electrochemical corrosion products [1,2].

Fig. 5.8a shows a BSE image at the steel–concrete interface. It can clearly be seen that the iron oxides include two layers with the clear interface. The oxide near the concrete is brighter and the oxide adjacent to the steel bar is relatively darker. The EDS analysis (Fig. 5.8b) of an analytical line

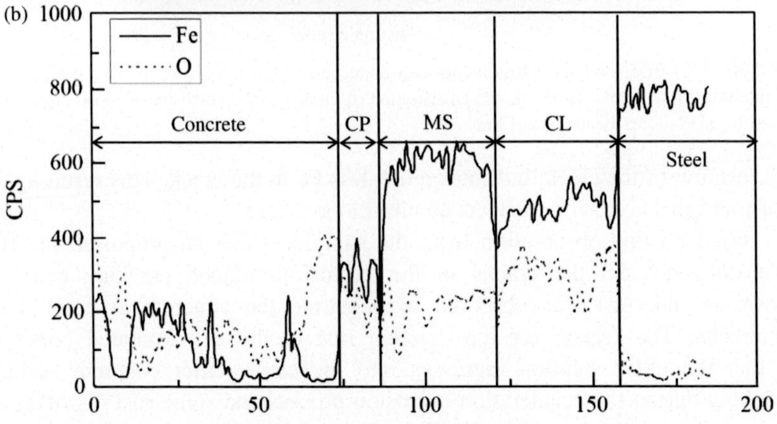

FIGURE 5.8 EDS analysis across the steel–concrete interface. (a) BSE image at the steel–concrete interface (*MS*, mill scale) and an analytical line across the interface. (b) The distribution of Fe and O across the steel–concrete interface analyzed by EDS along the analytical line.

FIGURE 5.9 Mill scale distribution at the steel–concrete interface.

in Fig. 5.8a shows five regions along the analytical line (ie, concrete, CP, MS, CL generated during the acceleration corrosion process, and steel, respectively). The mill scale is usually found near the outer edge of the rust layer. This suggests that movement of the mill scale has occurred due to the expansion and growth of the rust layer from the reaction zone at the steel boundary. This phenomenon is also found in another specimen, which is discussed in chapter "Rust Distribution in Corrosion-Induced Cracking Concrete."

In some areas, like the bottom left area of Fig. 5.9, the mill scale is completely oxidized and replaced by a thick layer of corrosion products. This is because the oxygen and water supplements are sufficient after a crack, which can be observed in the bottom left area, and the unstable FeO in the mill scale is further oxidized to the lower Fe-to-O atomic ratio corrosion products.

As mentioned, the mill scale is formed before corrosion initiation; it does not contribute to the volume expansion acting on the surrounding concrete cover. Therefore, the rust inducing the cracking of the concrete cover should not include the mill scale. Measurements were made at 10 locations of the remnants of dislocated mill scale. It was found that the thickness of the mill scale, T_m, remains stable, with a mean value of 34.5 μm. This value is used in Section 5.5 to subtract the thickness of the mill scale from the whole measured thickness of the rust layer across the perimeter of the rebar.

5.5 CORROSION LAYER THICKNESS AT SURFACE CRACKING OF CONCRETE COVER

5.5.1 At Outer Surface Cracking

Appropriate slices should be chosen carefully to measure the critical steel corrosion at concrete outer surface cracking. The crack of slice 1-1 penetrated

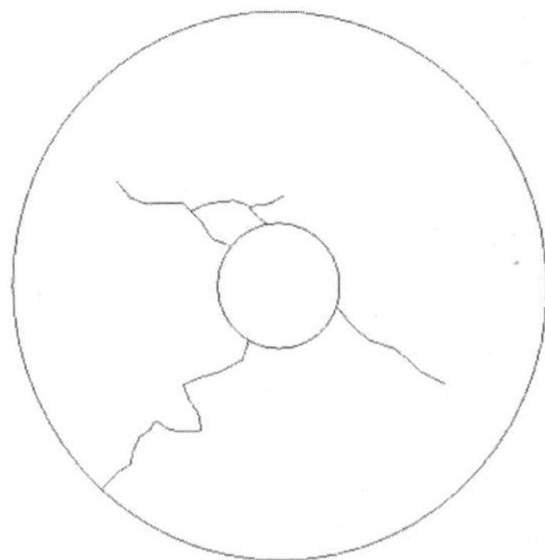

FIGURE 5.10 Crack pattern of slice 1-1.

the concrete cover, whereas the crack of slice 1-2 adjacent to slice 1-1 did not reach the concrete surface. Therefore, the steel corrosion of slice 1-1 was considered the critical steel corrosion at concrete outer surface cracking. Fig. 5.10 shows the crack pattern on the cross-section of slice 1-1.

The rust layer generated by the accelerated corrosion method appears featureless and uniform, but occasionally there are fluctuations at some locations. To obtain an accurate mean rust layer thickness, the area of the rust layer was measured by the digital microscope. Then, the average rust layer thickness was calculated by dividing the area of the rust layer by the perimeter of the rebar. It needs to be pointed out that the real corrosion layer thickness causing concrete cover cracking should be obtained from the total measured thickness, T_r, subtracting the mill scale thickness, T_m. Therefore, the corrosion layer thickness during concrete surface crack of slice 1-1 can be obtained as:

$$T_{CL}^{surface} = T_r - T_m = \frac{A_r}{2\pi R} - T_m = 86.8 - 34.5 = 52.3\ \mu m \tag{5.1}$$

The same method was adopted to obtain the critical rust layer thickness of specimen 2. Slice 2-1 was chosen for measurement; its crack pattern is shown in Fig. 5.11. The mean thickness of the corrosion layer of slice 2-1 is 45.6 μm.

For specimen 3, although a short crack exists at one side of the specimen, it had been observed that corrosion products leaked from this side during the accreted corrosion process (due to the unsuccessful surface treatment at this end of specimen 3). The steel corrosion measured at this region cannot represent the real situation without rust leakage. Therefore, specimen 3 is not taken into account while measuring the steel corrosion at surface cracking.

In this study, the measured thickness of the corrosion layer at concrete outer surface cracking is approximately 50 μm. However, if the mill scale is not subtracted, then the measured thickness of the rust layer is more than 80 μm, which obviously overestimates the critical thickness of the corrosion

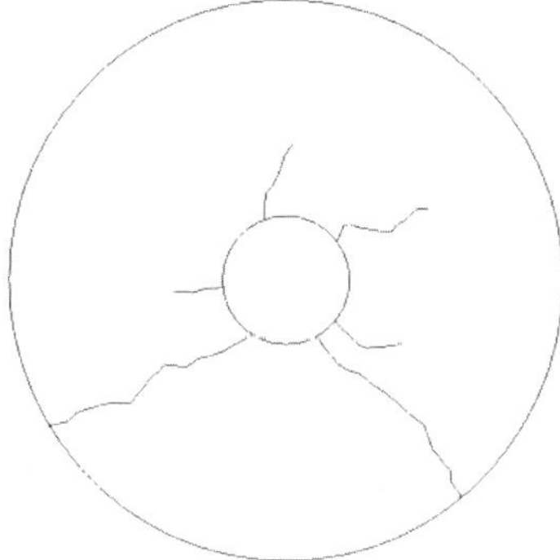

FIGURE 5.11 Crack pattern of slice 2-1.

layer at concrete surface cracking $T_{CL}^{surface}$. Therefore, to obtain the accurate $T_{CL}^{surface}$, the thickness of mill scale, T_m, must be subtracted from the measured thickness of the rust layer T_r.

5.5.2 At Inner Surface Cracking

The steel corrosion of the inner cracking of the concrete cover is difficult to measure because it is not possible to capture the inner cracking moment. To obtain the rust layer thickness at this moment, we measured the rust layer thickness and the radial crack length from the prepared slices that contain inner cracks in the concrete cover. The inner cracks here are the cracks that do not penetrate the concrete cover.

The crack patterns on the cross-section of the measured slices are shown in Fig. 5.12; the longest radial crack length and the corrosion layer thickness (using the same method stated in Section 5.5.1) are listed in the parentheses below each slice. Both sides of some slices (ie, slice 1-3, slice 2-2, slice 2-3, and slice 3-1) are observed. The two sides of one slice are labeled as (a) and (b).

The relation between the radial crack lengths of the inner cracks and the corrosion layer thickness is shown in Fig. 5.13. It can be seen that the crack lengths are linearly proportionally to the corrosion layer thickness as follows:

$$T_{CL} = 1.2542(R_c - R) + 15.161 \tag{5.2}$$

where T_{CL} is the corrosion layer thickness (μm), R_c is the radius at the interface between the cracked and intact cylinders (μm), R is the radius of the steel (μm), and $(R_c - R)$ represents the radial length of the longest inner crack. The determination coefficient, R^2, which can judge this fitting curve synthetically, is 0.93.

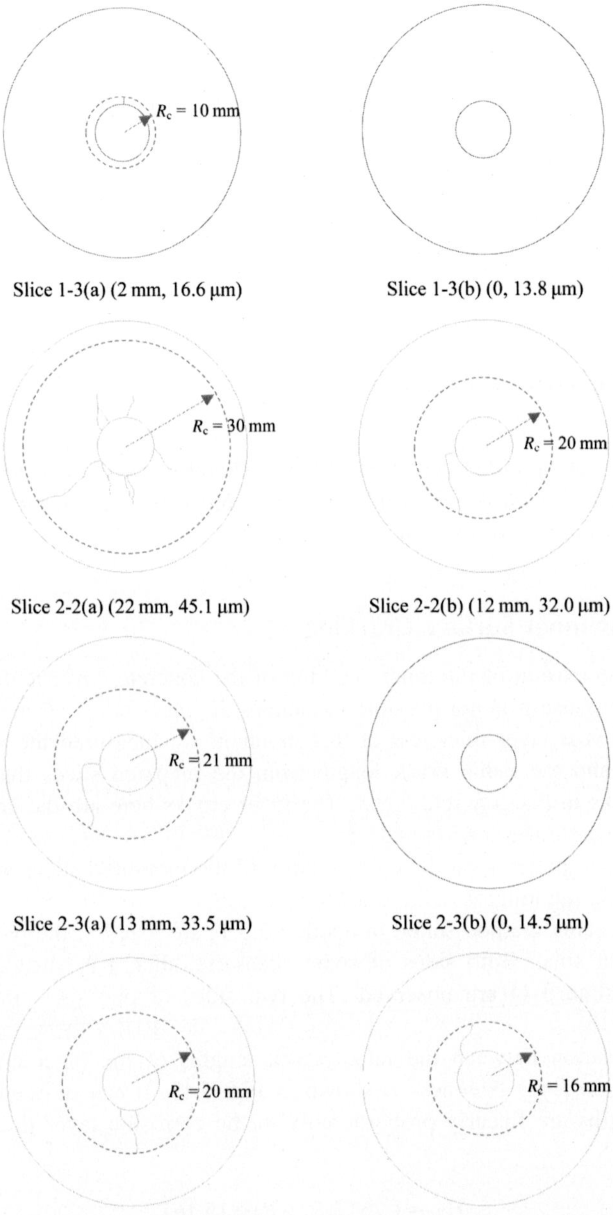

R_c = 10 mm

Slice 1-3(a) (2 mm, 16.6 μm)

Slice 1-3(b) (0, 13.8 μm)

R_c = 30 mm

Slice 2-2(a) (22 mm, 45.1 μm)

R_c = 20 mm

Slice 2-2(b) (12 mm, 32.0 μm)

R_c = 21 mm

Slice 2-3(a) (13 mm, 33.5 μm)

Slice 2-3(b) (0, 14.5 μm)

R_c = 20 mm

Slice 3-1(a) (12 mm, 33.5 μm)

R_c = 16 mm

Slice 3-1(b) (8 mm, 21.1 μm)

FIGURE 5.12 The crack patterns on the cross-section of the measured slices. The longest radial crack length and the corrosion layer thickness are listed below each slice.

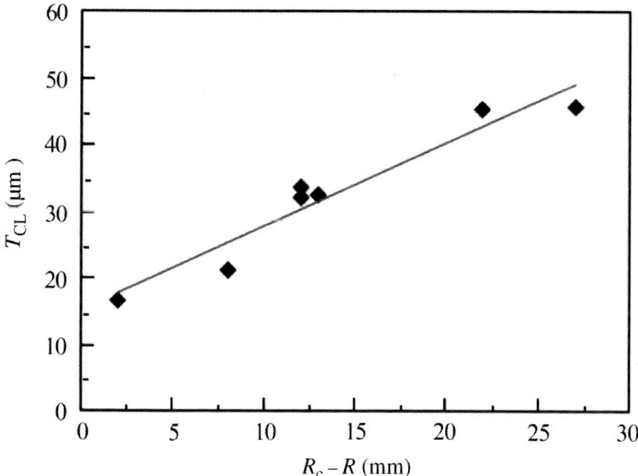

FIGURE 5.13 Relation between corrosion layer thickness and crack length.

Let $R_c - R = 0$; then, the corrosion layer thickness at inner surface cracking of the concrete cover can be obtained (ie, $T_{CL}^{inner} = 15.2$ μm).

As shown in Fig. 5.12, no crack was observed in the concrete cover of slice 1-3(b) and slice 2-3(b). The corrosion layer thicknesses of slice 1-3(b) and slice 2-3(b) are 13.8 and 14.5 μm, respectively, and both values are smaller than the regressed value of 15.2 μm. Thus, it is appropriate to use 15.2 μm as the critical corrosion layer thickness at inner surface cracking of the concrete cover.

It can be found that the thickness of mill scale T_m is mostly two-thirds of $T_{CL}^{surface}$ and two-times that of T_{CL}^{inner}. Therefore, the influence of mill scale should be considered in corrosion-induced concrete cracking model.

5.6 CONCLUSIONS

1. A uniform mill scale layer exists on the surface of the corroded steel bar. The mill scale moves from the steel boundary due to the expansion and growth of the corrosion layer. Because the mill scale is formed before corrosion initiation, it does not contribute to the volume expansion acting on the surrounding concrete cover. Therefore, the corrosion inducing the cracking of the concrete cover should not include mill scale.
2. The thickness of mill scale is comparable with the thickness of the corrosion layer at concrete surface cracking. Therefore, to obtain an accurate prediction of T_{CL}, the thickness of mill scale, T_m, must be subtracted from the measured thickness of the rust layer, T_r.
3. Rust does not fill the corrosion-induced cracks before concrete surface cracking in the electrochemically corroded reinforced concrete specimens.

REFERENCES

[1] Caré S, Nguyen QT, L'Hostis V, et al. Mechanical properties of the rust layer induced by impressed current method in reinforced mortar. Cem Concr Res 2008;38(8):1079–91.

[2] Wong HS, Zhao YX, Karimi AR, et al. On the penetration of corrosion products from reinforcing steel into concrete due to chloride-induced corrosion. Corros Sci 2010; 52(7):2469–80.

[3] Cornell RM, Schwertmann U. The iron oxides. Weinhaim, Germany: VHC Verlagsgesellschaft; 1996.

[4] Jaffer SJ, Hansson CM. Chloride-induced corrosion products of steel in cracked-concrete subjected to different loading conditions[Z]. Cem Concr Res 2009;39(2):116–25.

Chapter 6

Rust Distribution in Corrosion-Induced Cracking Concrete

Chapter Outline

6.1 INTRODUCTION

Based on the observations of the three cylindrical reinforced concrete specimens in chapter "Mill Scale and Corrosion Layer at Concrete Surface Cracking," it can be found that rust does not fill the cracks in corrosion-induced cracking before concrete surface cracking. However, in the electrochemically accelerated corrosion experiment, the high speed of steel corrosion may have resulted in insufficient time for corrosion products to fill the cracks; therefore, this result needs to be further proven.

The specimen investigated in this chapter, by contrast, had deteriorated in an artificial environment for 2 years, thus presenting more realistic rust distribution in corrosion-induced cracks. The rust distribution at the steel–concrete interface and in the corrosion-induced cracks was observed. Crack propagation induced by steel corrosion and rust development in cracks are also discussed in this chapter.

Steel Corrosion-Induced Concrete Cracking. DOI: http://dx.doi.org/10.1016/B978-0-12-809197-5.00006-2

6.2 EXPERIMENTAL PROGRAM

6.2.1 Reinforced Concrete Specimen

A reinforced concrete specimen with the dimensions shown in Fig. 6.1 was used in this study. Each specimen contained three ribbed bars with a nominal diameter of 16 mm and a 20 mm cover spacing. The rebars were used as received, and no efforts were made to remove the existing mill scale. A low cover depth of 20 mm was adopted to reduce the time for corrosion initiation.

The concrete used in the panel was a ternary blended mixture that contained slag and fly ash at 40 wt.% and 30 wt.% replacement levels, respectively. The mixture proportions used are reported in Table 6.1. The water-to-binder ratio was 0.345. The concrete contained a commercial calcium nitrite–based corrosion inhibitor. The gravel and sand were siliceous aggregates at 20 mm and 5 mm maximum size, respectively. The 28 day compressive strength of the concrete was 56 MPa as measured on 150 mm cubes.

6.2.2 Curing and Exposure History

The specimen was covered in damp hessian immediately after casting and sprayed with water once every day for 2 weeks. Thereafter, the panel was moved into the laboratory and dried at room temperature

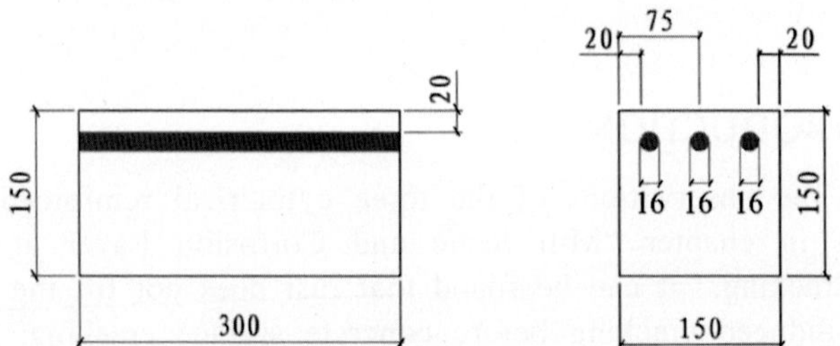

FIGURE 6.1 Schematic of the reinforced concrete specimen (dimensions are in mm).

TABLE 6.1 Mixture Composition of Concrete Specimens (kg/m^3)

Cement	Blast-Furnace Slag	Fly ash	Sand	Aggregate	Water	*w/c*	Water-Reducing Admixtures	Corrosion Inhibitor
126	168	126	735	1068	145	0.345	5.04	8.4

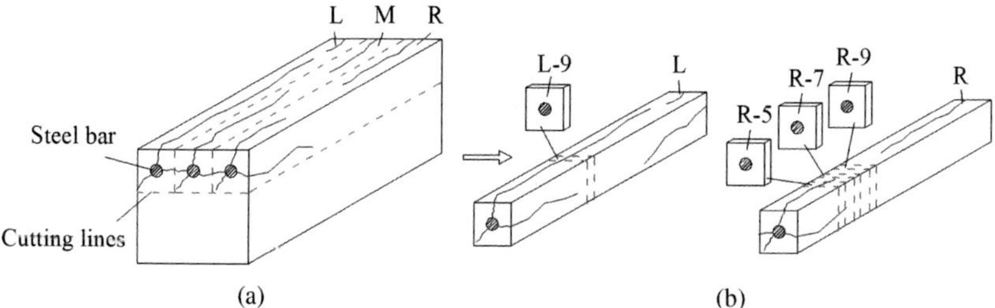

FIGURE 6.2 Schematic diagrams of the cut specimen. (a) Specimen. (b) The cut panels and slices.

(approximately 25°C) for several weeks. The side and bottom surfaces of each panel were surface-treated with epoxy resin and a polyurethane coating to ensure that chloride predominantly penetrated through the top cover, with minimal penetration through the other faces of the panel.

The concrete panel was then subjected to alternate wetting and drying cycles in a walk-in environmental chamber. Each test cycle lasted for 3 days and consisted of spraying the blocks with 3.53 wt.% sodium chloride solution for 4 h and subsequently allowing them to dry at 40°C for the remainder of the time to complete one full cycle.

After 2 years of exposure, longitudinal cracks were observed on the top face of the specimen, running approximately parallel to the reinforcing bar. However, the longitudinal cracks did not run across the entire top surface of the specimen; some parts at the center of the top surface did not show cracks, as shown in Fig. 6.2. The widths of the cracks in the areas near both sides of the specimens were obviously wider than those in the middle part. Brownish-red rust stains were observed on the external side surfaces of the specimen. The specimen was then removed from the environmental chamber for further testing.

6.2.3 Sample Preparation

The selected specimen was cast into a low-viscosity epoxy resin to minimize any artificial damage that may have been incurred during the sample preparation process for microscopy. Then, it was carefully cut to extract the corner and middle rebars while maintaining their surrounding concrete intact, as shown in Fig. 6.2. The cut panels were labeled L, M, and R, representing the sections containing the left-corner rebar, middle rebar, and right-corner rebar, respectively.

Each panel was then sectioned sequentially, starting from the side face, to produce a series of 10-mm-thick cross-section slices. Cutting was performed using an abrasive cutter and diamond blade suitable for hard brittle materials. An example of a slice is shown in Fig. 6.3a. The locations of the

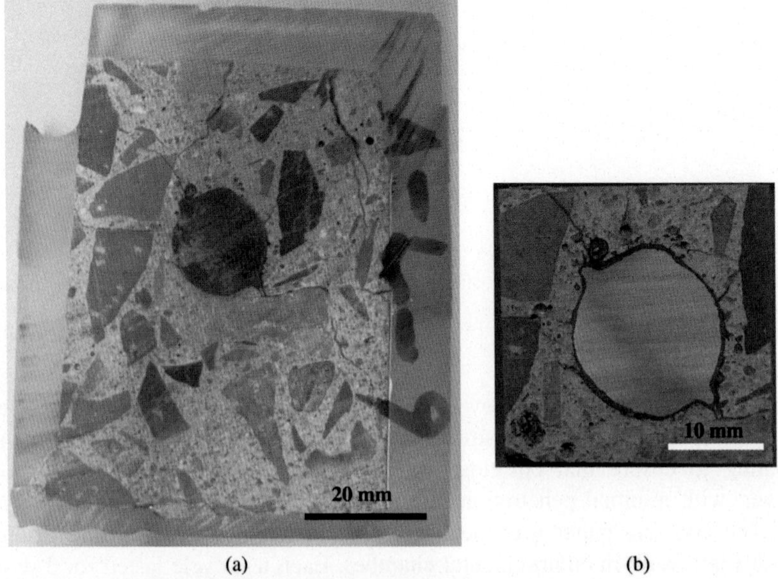

FIGURE 6.3 Sample preparation for SEM. (a) Slice L-9. (b) Sample for SEM.

slices are indicated in Fig. 6.2; for example, "R-7" represents the seventh slice from the side face of this panel containing the right-corner rebar. The samples were further prepared for scanning electron microscopy (SEM) observation by trimming slices to smaller blocks, including the complete steel–concrete interface and the cracks (Fig. 6.3b)

To prevent further corrosion, these slices were kept in a dry environment (relative humidity less than 30%) before observation. Polishing at each stage was performed using different specifications of abrasive paper.

6.2.4 Observation and Measurements

The observation and measurements of the rust layer and the steel–concrete interface are the same as stated in Section 5.2.4. Approximately 80 points evenly distributed around the rebar were observed to cover the entire steel–concrete interface, as illustrated in Fig. 6.4.

The slices from panels L, M, and R were all investigated to study the rust distribution. According to the colors and locations of the rust, the distribution of different types of rust was observed by digital microscopy. The corrosion-induced cracks were also observed using SEM in the backscattered electron (BSE) mode. The content of Fe was analyzed by energy spectrum analysis

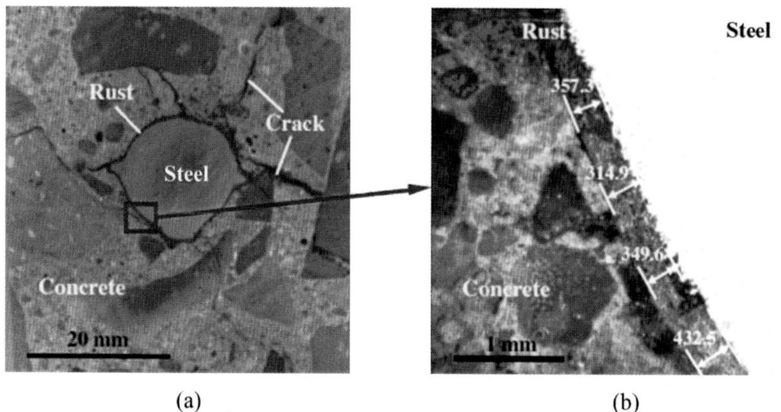

FIGURE 6.4 Measurement of the thickness of the rust layer accumulated at the rebar–concrete interface. (a) Field of view: 50 × 52 mm. (b) Field of view: 3.74 × 3.68 mm.

(EDS) to investigate rust distributions in cracks. Based on digital microscopy and the SEM observations, trends in the rust distribution in the corrosion-induced cracking concrete specimen were then analyzed.

6.3 RUST DISTRIBUTIONS AT THE STEEL–CONCRETE INTERFACES

Fig. 6.5a shows a BSE image at a steel–concrete interface observed in sample R-5, which demonstrates the typical rust distribution at the steel–concrete interface and in concrete according to observation. The distributions of Fe and O across the steel–concrete interface were analyzed by EDS along an analytical line, as shown in Fig. 6.5a. In Fig. 6.5b, the horizontal axis represents the distance from the starting point of the analytical line, and the vertical axis represents the counts per second of photoelectrons, reflecting the contents of Fe and O, respectively. Like the observation of the electrochemically corrosion specimens reported in chapter "Mill Scale and Corrosion Layer at Concrete Surface Cracking", the EDS analysis in this investigation also indicates five regions along the analytical line: concrete, corrosion products-filled paste (CP), mill scale (MS), corrosion layer (CL), and steel, respectively. The formation of these five regions is discussed here.

In the presence of chloride ions in concrete, the reaction between Fe^{2+} formed by the oxidation of iron in the anodic region (steel) and OH^- formed by oxygen in the cathodic region (concrete) is believed to involve soluble intermediary "green complexes" [1], which break down once they encounter OH^- and undergo further oxidation to form solid precipitates. The possible

(a)

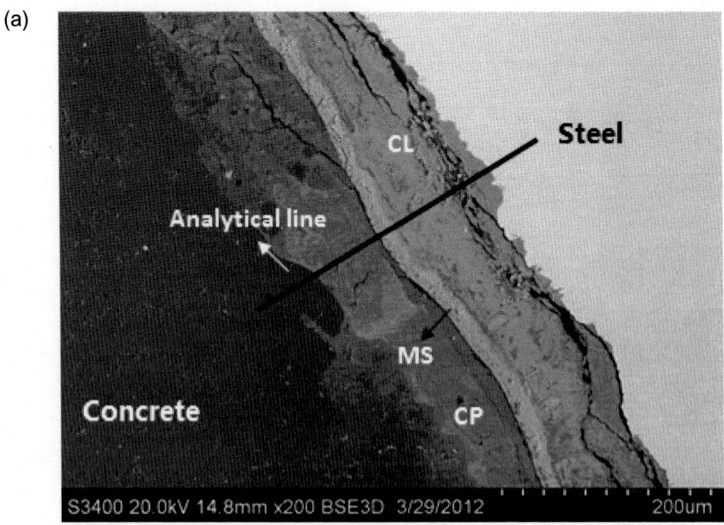

(b)

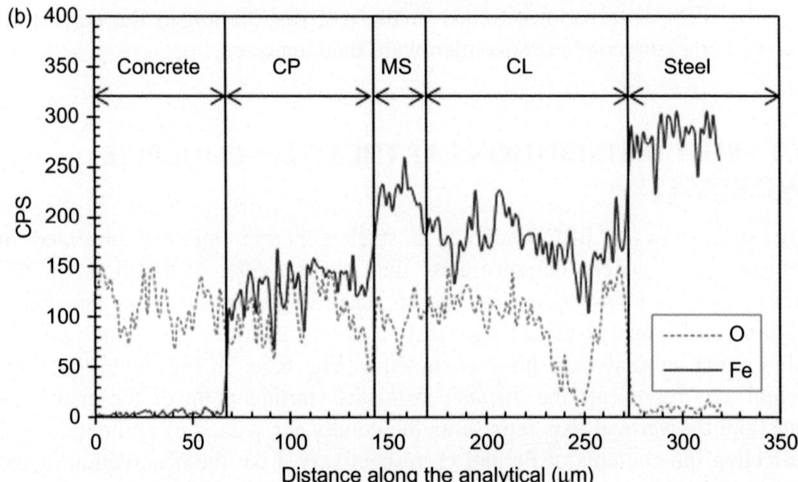

FIGURE 6.5 Rust distributions at the steel–concrete interface in sample R-5. (a) BSE image at the steel–concrete interface (CP, corrosion products-filled paste; MS, mill scale; CL, corrosion layer) and an analytical line across the interface. (b) The distributions of Fe and O across the steel–concrete interface analyzed by EDS along the analytical line.

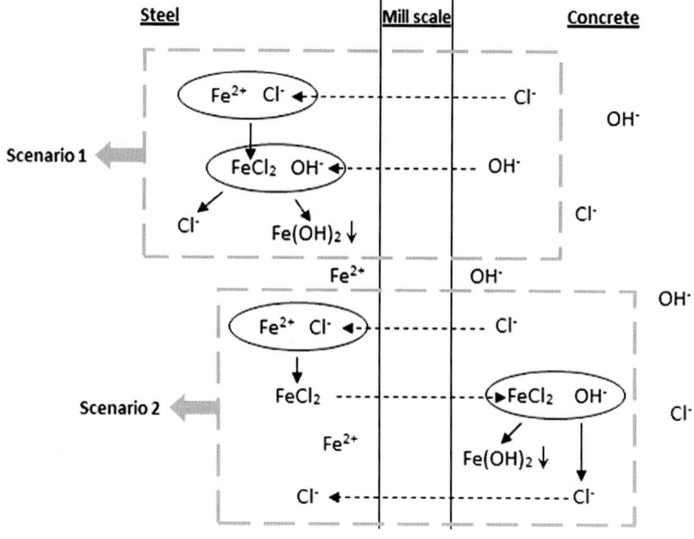

FIGURE 6.6 Schematic of ion migration and reaction during steel corrosion in the presence of chloride ions in concrete.

reactions during steel corrosion are shown (by assuming that the green complexes are $FeCl_2$ as stated previously [2]):

$$\text{Anode: } 2Fe \rightarrow 2Fe^{2+} + 4e^{-} \tag{6.1}$$

$$2Fe^{2+} + 4Cl^{-} \rightarrow 2FeCl_2 \tag{6.2}$$

$$\text{Cathode: } O_2 + 2H_2O + 4e^{-} \rightarrow 4OH^{-} \tag{6.3}$$

$$\text{Overall: } 2Fe + O_2 + 2H_2O \rightarrow 2Fe(OH)_2 \tag{6.4}$$

The mill scale of a rebar embedded in concrete was found to be porous [3], allowing soluble species involved in the aforementioned reaction to move through it from the anodic to cathodic regions or in the opposite direction. The ionic reaction and migration process can be simplified, as represented in Fig. 6.6. In scenario 1, OH^- formed in the cathodic region (concrete) penetrates through the mill scale into the surface of the steel. When OH^- encounters $FeCl_2$, $FeCl_2$ breaks down and is further oxidized to $Fe(OH)_2$. This scenario briefly describes the process by which the steel corrosion accumulates between the steel and mill scale, forming CL. However, in scenario 2, as shown in Fig. 6.6, the soluble intermediary green complex $FeCl_2$ moves through the porous mill scale away from the steel surface to the adjacent concrete, and then $FeCl_2$ and OH^- react and form $Fe(OH)_2$, which is the same reaction as in scenario 1. The precipitate $Fe(OH)_2$ fills in the porous cement paste, resulting in the formation of CP. It needs to be noted that the chloride ions take part in the formation of these different layers in this study, For other environments that can cause rebar to corrode, such as carbonation, Fe^{2+} (or Fe^{3+}) can also move from the surface of rebar and react with OH^- and form $Fe(OH)_2$ without the appurtenance of Cl^-.

The mill scale was preserved well after the growth of the CL because the iron (Fe) in steel is more easily oxidized than ferrous iron in iron oxides in mill scale under the low-oxygen conditions in concrete. Therefore, generally speaking, the mill scale of the rebar cannot be further oxidized easily before surface cracking of the concrete cover.

6.4 DISTRIBUTION OF THE CORROSION PRODUCTS-FILLED PASTE

Samples L-9, R-5, and R-7, spanning a range of degrees of corrosion and damage, were analyzed using SEM in BSE mode. The steel–concrete interfaces were observed at approximately 80 points that were evenly distributed around the rebar to cover the entire perimeter. The thicknesses of the CL and the CP were measured at every point observed.

Although the data collected by the observations and measurements are scattered, the statistical results of the data from all samples observed display some regularity when these scattered data are graded into several groups, as shown in Fig. 6.7. The data for each group in Fig. 6.7 are reported in Table 6.2. It could be noted that the data for the first group are the biggest. This is because the corrosion products normally accumulate on the surface of the steel bar facing the concrete surface; very little corrosion is observed on the opposite side of the steel bar for each slice. It also needs to be mentioned that the thickness of the CP in the same group varies widely.

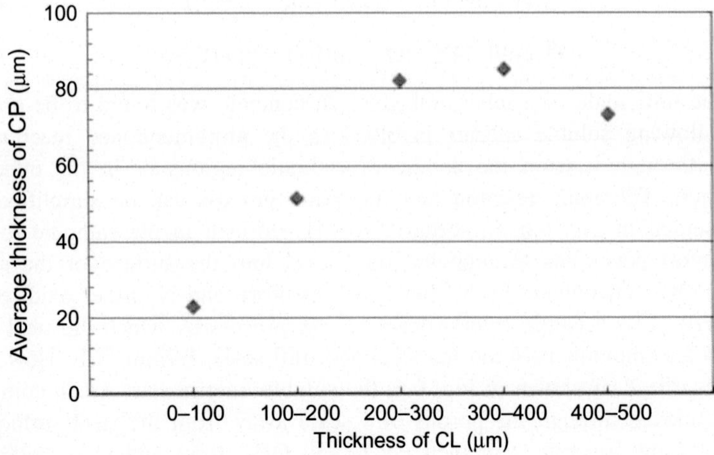

FIGURE 6.7 Average thickness of the corrosion products-filled paste (CP) for different thicknesses of the corrosion layer (CL).

TABLE 6.2 Data for Each Group in Fig. 6.7

CL thickness, μm	0–100	100–200	200–300	300–400	400–500
Number of tested data	105	37	29	18	13

There are many factors influencing the penetration progress of rust into concrete, such as the location of the aggregate and pore distribution in cement. Considering the great heterogeneity of concrete, the average value of thickness of the CP is introduced in Fig. 6.7.

The abscissa in Fig. 6.7 illustrates the range of the CL thickness in each group, and the ordinate is the average thickness of the CP in each group. The result reveals that the average thickness of the CP increases with the growth of the CL; this tendency does not change until the thickness of CL exceeds 300 μm. The regularity differs from the assumption of the widely accepted three-stage model [4]. The three-stage model [4] assumes that the corrosion products will never create expansive pressure on the surrounding concrete before the porous zone around the steel–concrete interface is fully filled. The results here, however, reveal that the penetration of corrosion products into the porous zone of concrete and the formation of the CL at the steel–concrete interface might, in fact, proceed simultaneously after the initiation of steel corrosion. The mechanism of this fact has been discussed in Section 6.3.

In addition, compared to previous studies [5,6], the value of the average thickness of CP obtained in this research is smaller, possibly because the corrosion degree of the samples observed in previous studies is higher than that in this research. This factor may have an influence on the distribution of rust in concrete.

It should be recognized that, because the samples involved in this study are limited, more experiments should be performed with more samples spanning a larger range of degrees of corrosion. Therefore, in chapter "Development of the Corrosion Products-Filled Paste at the Steel/Concrete Interface," four type of concrete specimens are specifically investigated to study the relationship of the thicknesses of the CP and the thickness of CL. For more detailed information, please refer to chapter "Development of the Corrosion Products-Filled Paste at the Steel/Concrete Interface."

6.5 RUST DISTRIBUTION IN CORROSION-INDUCED CRACKS

6.5.1 Rust Distribution in Cracks by Digital Microscope

The cracks in this study are classified into two types: external cracks and internal cracks. The cracks that penetrate the concrete cover are defined as the external cracks, whereas the cracks that cannot propagate to the surface

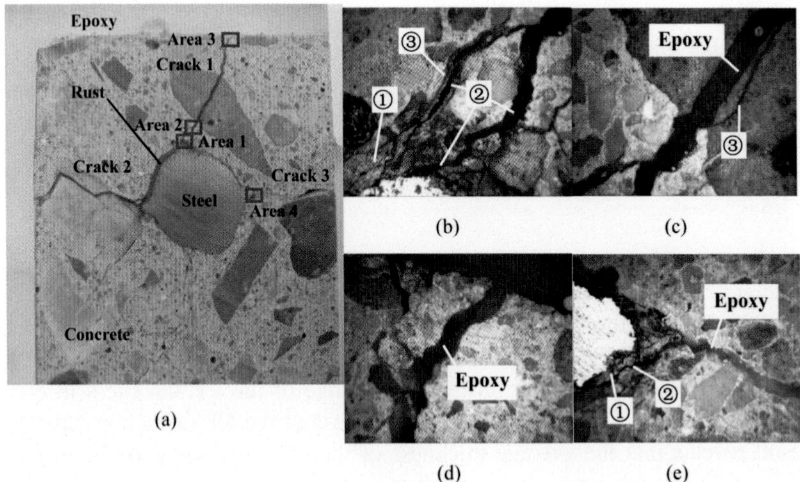

FIGURE 6.8 Rust distribution in slice R-6. (a) Slice R-6. (b) Area 1. (c) Area 2. (d) Area 3. (e) Area 4.

of the concrete cover, including those propagating inward rather than outward to the concrete surface and the ones developing between two steel bars, are defined as the internal cracks.

Slice R-6 (located in the middle of the panel R) is taken as an example to introduce the rust distribution in corrosion-induced cracks. A cross-section of this slice is shown in Fig. 6.8a. The top surface of the slice was exposed to the chloride ingress; the left side is the polyurethane-treated surface, and the right side and the bottom are the cutting surfaces. There are three corrosion-induced cracks in this slice, cracks 1–3, as shown in Fig. 6.8a. Typical images showing the local distribution of rust in these cracks are shown in Fig. 6.8b–e. The size of the observed area is 4718 μm × 3767 μm.

6.5.1.1 External Cracks

Cracks 1 and 2 both propagate to the surface of the concrete cover because the rust distributions in these two cracks are similar. Here, crack 1 is used as an example to illustrate the characteristics of the rust in the external cracks. Area 1 is close to the steel bar. As shown in Fig. 6.8b, the corrosion products clearly fill the cracks. Three types of corrosion products can be distinguished: rust ① at the steel/concrete interface; rust ② filling the corrosion-induced cracks; and rust ③ lining the edges of the corrosion-induced cracks and the surrounding microcracks.

Rust ① appears black with a silver metallic luster. Rust ① adheres to the steel surface and induces expansive pressure on the surrounding concrete cover, which is the direct cause of concrete surface cracking. Due to the restriction of the surrounding concrete, rust ① appears very dense.

Rust ② has a dark brown appearance. Rust ② fills the cracks of the rust layer in addition to the cracks in the concrete cover, as shown in Fig. 6.8b. As rust ② fills the empty space of the existing cracks, it is not as subjected to pressure as rust ① is; therefore, it is not as dense as rust ①.

Rust ③ is reddish brown and is observed along the edges of the corrosion-induced cracks and the concrete microcracks around the steel bar and cracks. It occurs because the outer solution enters the cracks after they penetrate the concrete cover; some rust dissolves in the solution and contaminates the concrete it contacts. When the solution dries during the drying cycles, the rust adheres to the edge of the cracks. With rich oxygen supplementation, the quantivalence of rust compounds became higher; therefore, rust ③ appears reddish.

In area 2, which is connected to area 1 as shown in Fig. 6.8c, the space of the upper-right crack is empty. The edges of the cracks at the bottom left of the area are stained by rust ③, which is reddish brown in color. There is no rust observed between areas 2 and 3.

Crack 1 penetrate the rust layer as shown in Fig. 6.8b. However, it is not commonly found. More images show that the rust layer does not crack when the concrete cracks. For instance, Fig. 6.9 shows that the cracks at the steel–concrete interface area in slice M-14 are clearly different from those observed in area 1 of slice R-6. The same results were observed in the other slices with cracks at the steel–concrete interface. This illustrates that concrete cover cracking occurs earlier than rust layer cracking.

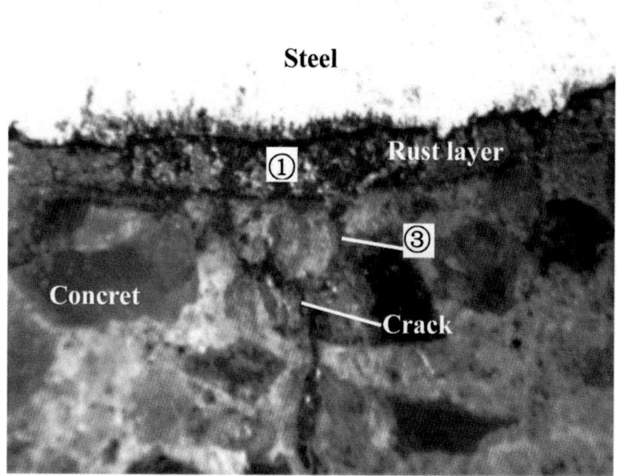

FIGURE 6.9 Crack at the steel–concrete interface of slice M-14 (6897 μm × 6155 μm).

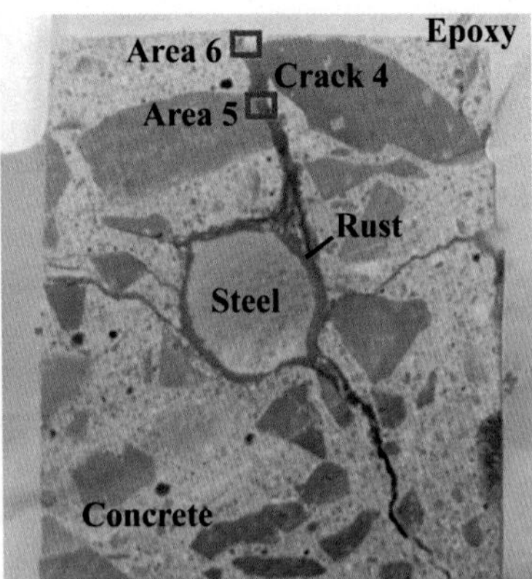

FIGURE 6.10 Slice L-4 with the more severe corroded steel bar.

6.5.1.2 *Internal Cracks*

In slice R-6, crack 3 is a crack generated between two steel bars (ie, an internal crack). The rust distribution in this kind of crack is different from that in external cracks 1 and 2. It can be seen from area 4 in Fig. 6.8e that although the crack width is approximately 60 μm and the rust layer has cracked, very little rust can be observed in the cracks. This is because the outer solution can barely penetrate into the internal cracks to solvate and carry the rust ions. Therefore, the ingress of the outer solution is the reason why the rust distribution between external and internal cracks differs.

6.5.1.3 *Influence of Steel Corrosion on Rust Distribution*

Observation of all of the slices shows that the range of rust distribution becomes wider as steel corrosion increases. Slice L-4 (located in the middle of the panel L), for example, shown in Fig. 6.10, features severe cracking and a surface crack width of 2.865 mm for crack 4. The rust can be observed lining the surface of the crack in area 5, which is far from the steel bar but close to the concrete surface. Rust can even be found on the edges of the crack in area 6, which is on the surface of the concrete cover. This is understandable because more severely corroded rebar produces more corrosion products. Moreover, the slices with more severely corroded steel bar normally exhibit wider external cracks, leading to the easier ingress and circulation of the outer solution during the wetting and drying cycles. These findings also suggest that the rust could be carried farther away from the rebar.

6.5.2 Rust Filling in Cracks by SEM

Fig. 6.11a illustrates the BSE images of rust distribution in corrosion-induced cracks obtained from sample L-9, in which the cracks penetrated the

(a)

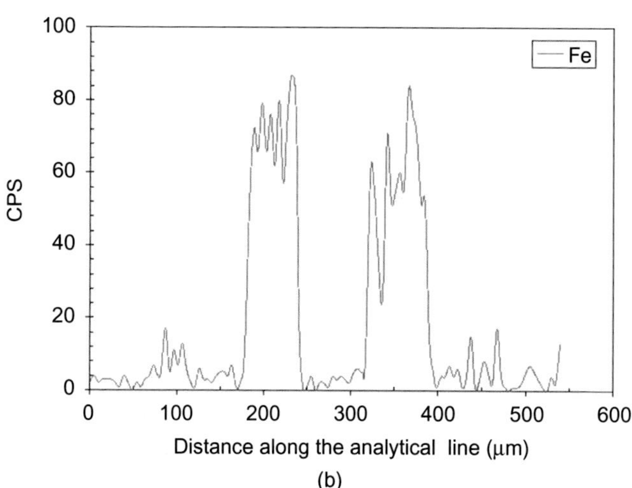

(b)

FIGURE 6.11 Rust distributed in a crack penetrating the concrete cover in sample L-9. (a) BSE image of the corrosion-induced crack and an analytical line across the crack. (b) The distribution of Fe across the crack analyzed by EDS along the analytical line.

concrete cover, and Fig. 6.12a shows the situation of sample R-7 with only the inner cracks. Fig. 6.11a shows that the concrete on both sides of the crack is brighter, corresponding to the peak Fe content in the EDS analysis in Fig. 6.11b, indicating that corrosion products penetrate the concrete adjacent to the cracks but that the crack itself is not filled. This phenomenon is the same as the observation of the electrochemically corrosion specimens reported in chapter "Mill Scale and Corrosion Layer at Concrete Surface

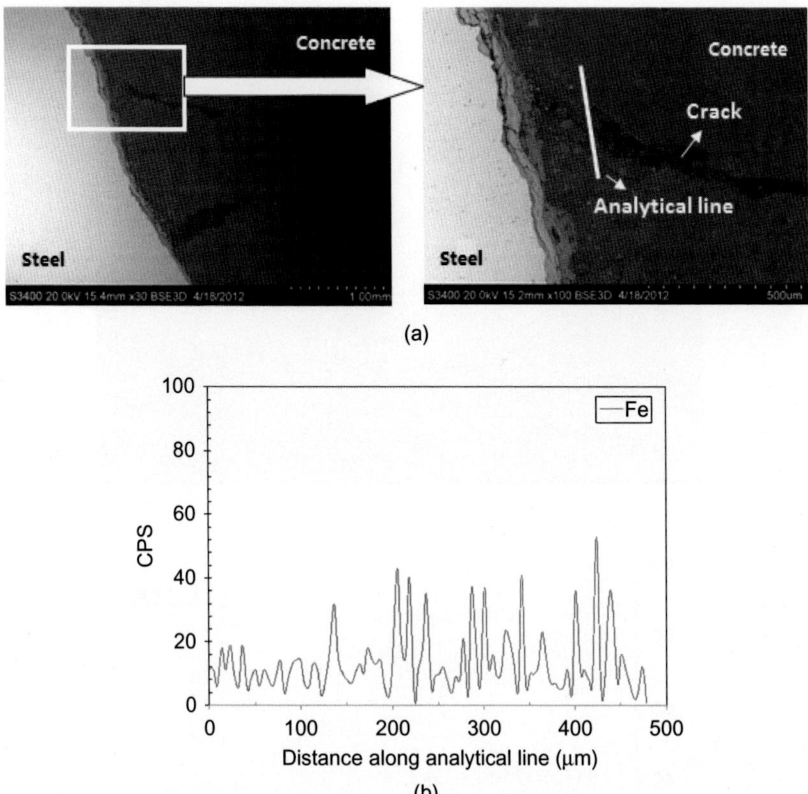

FIGURE 6.12 Rust distributed in an inner crack in sample R-7. (a) BSE image of the corrosion-induced crack and an analytical line across the crack. (b) The distribution of Fe across the crack analyzed by EDS along the analytical line.

Cracking." For the situation of the inner crack in sample R-7, the BSE image in Fig. 6.12a and the EDS analysis in Fig. 6.12b show no rust filling the crack and no rust lining the surface of the cracks because the concentration of Fe remains low along the entire analysis line across the inner crack.

Based on the observations of samples in this work, it can be proposed that rust does not fill the corrosion-induced cracks in concrete that has deteriorated in an artificial environment. Comparing Fig. 6.11 with Fig. 6.12, it can be seen that the rust does not penetrate the concrete at the sides of a crack until the crack reaches the concrete surface. In fact, only after the crack penetrates the concrete cover during the acceleration corrosion process can the outer solution ingress into the crack. Then, some corrosion products dissolve in the solution and are carried away from the rebar by the solution, thereby lining and penetrating the edges of cracks.

6.5.3 Discussion of Rust Filling Corrosion-Induced Cracks

Before steel–concrete interface cracking, rust deposits on the interface gradually. This rust layer is compressed by the surrounding concrete and is therefore dense. As steel corrosion increases, the concrete at the steel–concrete interface cracks first, whereas the rust remains deposited on the steel–concrete interface due to the protection of the dense rust layer. Only after the rust layer cracks is the rust able to fill the cracks.

This study shows that for external cracks with small crack widths, rust can rarely be found within the cracks. The test results again support that rust does not fill cracks before concrete surface cracking, as found in chapter "Mill Scale and Corrosion Layer at Concrete Surface Cracking."

After concrete surface cracking, the outer solution penetrates into the concrete through the cracks. Some rust dissolves in the solution, and the circulation of the solution removes the rust from the steel bar. When the specimen dries during the wetting and drying cycles, the rust remains in the cracks. Therefore, except for rust filling the cracks close to the rebar, rust can also be observed to absorb into the edges of the cracks far from the rebar.

Rust can also move into the internal cracks after the cracking of the rust layer. However, because the outer solution cannot penetrate these cracks, there is no rust found lining the edges of the cracks without circulation of the solution.

6.6 RUST DEVELOPMENT IN CONCRETE CRACKS

A schematic diagram is proposed to describe rust development during the corrosion-induced cracking process, as shown in Fig. 6.13.

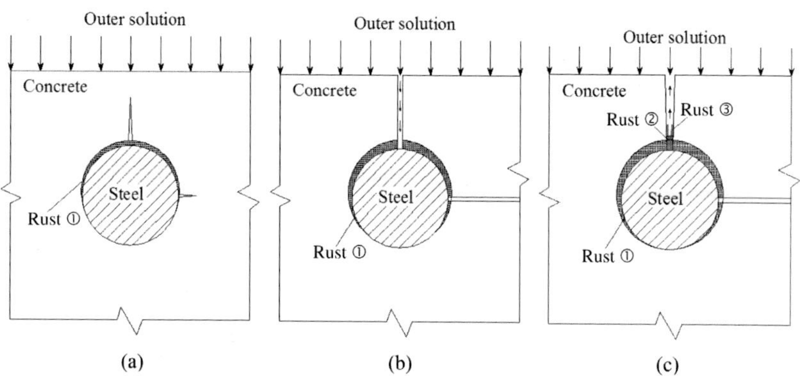

FIGURE 6.13 Schematic diagram of crack propagation and rust development. (a) Before surface cracking. (b) Surface cracking. (c) After surface cracking.

For external cracks, as shown in Fig. 6.13, rust develops as follows:

1. The concrete at the steel–concrete interface cracks and the rust (rust ①) still accumulates at the steel–concrete interface, leading to the propagation of the cracks in the concrete cover as shown in Fig. 6.13a.
2. The dense rust layer cracks, and the cracks propagate to the concrete surface in a relatively short amount of time; the rust filling the cracks can be ignored before surface cracking.
3. After concrete surface cracking, the outer solution ingresses into the cracks and steel corrosion is accelerated, as shown in Fig. 6.13b. If there is no solution on the surface of the sample, then the moisture (H_2O) and oxygen (O_2) will ingress through the cracks and accelerate the steel corrosion as well.
4. As the steel corrosion grows, rust ② fills the cracks. The rust is carried away from the rebar by the solution, lining the edges of the cracks (ie, rust ③), as shown in Fig. 6.13c.

Rust ② fills the internal cracks after the cracking of the rust layer at a very slow speed, and no rust lines the edges of the cracks.

According to this discussion, it can be assumed that filling corrosion-induced cracks with rust does not need to be considered when studying the concrete cracking process before surface cracking. Thus, the rust causing concrete cracking develops via only two processes: the rust migrates to the pores at the steel–concrete interface and the rust creates stress in the concrete cover.

6.7 CONCLUSIONS

1. There are five regions around the interface of steel and concrete: concrete, CP, mill scale, the CL, and steel, respectively.
2. The results of this study show that the penetration of corrosion products into the porous zone of concrete and the formation of a CL at the steel–concrete interface may proceed simultaneously after the initiation of steel corrosion.
3. Rust does not penetrate into the corrosion-induced cracks before the cracks reach the concrete surface. Thus, the rust filling the corrosion-induced cracks doses not need to be considered in corrosion-induced concrete surface cracking model.
4. After surface cracking, rust fills the cracks and develops in the external cracks much more quickly than in the internal cracks. Moreover, rust lines the edges of the external cracks due to the circulation of the outer solution.

REFERENCES

[1] Sagoe-Crentsil KK, Glasser FP. "Green rust", iron solubility and the role of chloride in the corrosion of steel at high pH. Cement Concrete Res 1993;23(4):785–91.

[2] Shi HS, Guo XL, Zhang H. Influence of chloride anion on the corrosion of steel bar in concrete. Cem Tech 2009;5:21–5.

[3] Jaffer SJ, Hansson CM. Chloride-induced corrosion products of steel in cracked-concrete subjected to different loading conditions. Cement Concrete Res 2009;39(2):116–25.

[4] Liu Y, Weyers RE. Modeling the time-to-corrosion cracking in chloride contaminated reinforced concrete structures. ACI Mater J 1998;95(6):675–81.

[5] Wong HS, Zhao YX, Karimi AR, et al. On the penetration of corrosion products from reinforcing steel into concrete due to chloride-induced corrosion. Corrosion Sci 2010; 52(7):2469–80.

[6] Michel A, Pease BJ, Geiker MR, et al. Monitoring reinforcement corrosion and corrosion-induced cracking using non-destructive x-ray attenuation measurements. Cement Concrete Res 2011;41(11):1085–94.

Chapter 7

Nonuniform Distribution of Rust Layer Around Steel Bar in Concrete

Chapter Outline

7.1 INTRODUCTION

A considerable amount of research has been undertaken regarding the cracking process of the concrete cover due to reinforcement corrosion. However, the analytical models [1–11] usually assume a uniform layer of corrosion products around the rebar's circumference. A number of experimental studies have been performed to investigate the radial loss of steel bar at the concrete surface cracking by the electro-chemical method to accelerate steel corrosion in the concrete specimens [12–17]. Previous theoretical and experimental studies are all based on the assumption of uniform corrosion; however, this is rarely the case in practice. The nonuniform corrosion around the rebar perimeter is the real situation.

The adoption of uniform corrosion in previous work is mainly due to three reasons. In the experimental process, the electro-chemical method is a widely used and easy way to accrete steel corrosion in concrete, although it results in uniform corrosion if the chloride solution is not introduced into the concrete in advance. For analytical studies, it simplifies the modeling process.

Steel Corrosion-Induced Concrete Cracking. DOI: http://dx.doi.org/10.1016/B978-0-12-809197-5.00007-4

The third reason is that the corrosion products at the steel–concrete interface have been studied previously [18–21]; however, there is little reliable information with which to characterize the actual nonuniform formation and expansion of corrosion products from the rebar surface [22–24]. It must be noted that although the variation in corrosion rate is the fundamental cause of the variability in the prediction of the time to corrosion-induced cracking, the scenario of the uniform or nonuniform corrosion of reinforcement obviously has a different effect on the prediction. Hence, to obtain a reasonably accurate prediction of the time to corrosion-induced cover cracking, not only the quantity of rust formed but also the variability in the thickness of the rust layer deposited on the circumference of a steel bar must be considered when formulating a mathematical model.

We measured and investigated the thickness of a nonuniform rust layer around the steel bar in a concrete specimen that is from the same batch of specimens used in chapter "Rust Distribution in Corrosion-Induced Cracking Concrete." The experimental program, including the specimens and their history, the sample preparation, and the rust layer measurement are also the same as described in chapter "Rust Distribution in Corrosion-Induced Cracking Concrete."

7.2 STEEL CORROSION AND CORROSION-INDUCED CRACKS

The cover-penetrating cracks induced by steel corrosion can be observed in most samples. The amount of corrosion, defined as the measured area of corrosion products expressed as the percentage of the original rebar cross-sectional area, ranges from 3% to 9%. Slices from the corner rebar are slightly more damaged compared to those from the middle rebar, given the same amount of corrosion. Corrosion is more severe on slices taken near the side compared to those farther away, as illustrated by Fig. 7.1. This is because steel corrosion at the front damages the surface treatment of the front during the exposure period; it is easier for chloride ions to ingress from this side, resulting in high corrosion

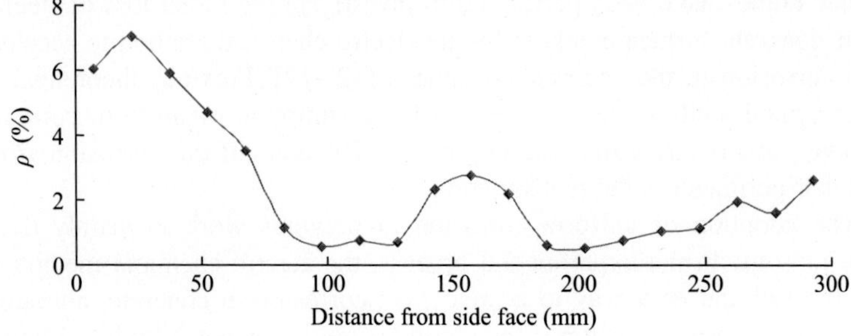

FIGURE 7.1 Steel corrosion varies with the distance to the front of specimen R.

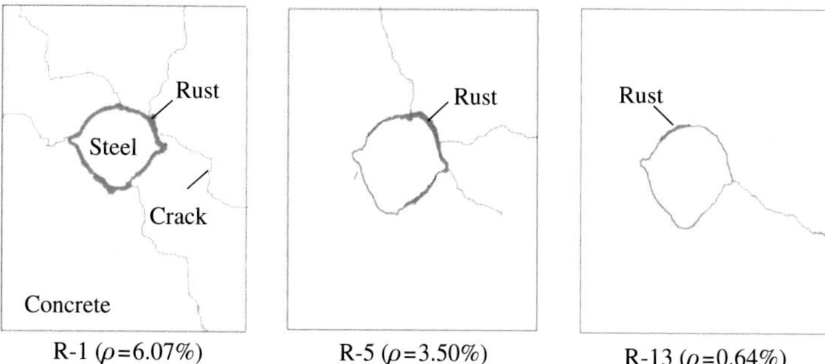

FIGURE 7.2 Typical cracks and rust layer of slices from specimen R.

at the front of the concrete penal. It is also clearly observed that the amount of corrosion products at the rebar–concrete interface is not uniform around the rebar, but rather accumulates mainly in the upper half, where it is closest to the exposed top cover.

Typical corrosion-induced cracks are shown in Fig. 7.2. It can be observed that the quantity of corrosion-induced cracks increases with the growth of steel corrosion.

7.3 GAUSSIAN MODEL TO DESCRIBE THE NONUNIFORM RUST LAYER

Fig. 7.3 shows the variation in the measured rust layer thickness around the rebar circumference for the five typical samples. The results are plotted against polar coordinates where the origin ($\theta = 0$) is defined according to the schematic shown in Fig. 7.4 for the corner and middle rebars. The results show large scattering in measured thickness, but some trends are apparent. As the sample becomes more corroded, the spread and thickness of the rust layer increase. The largest recorded thickness is approximately 1 mm.

The data were fitted using several equations, and it was found that an equation based on a Gaussian function is suitable to describe the rust layer thickness around the rebar for the case of nonuniform corrosion:

$$T_r = \frac{\lambda_1}{\lambda_2 \cdot \sqrt{2\pi}} \cdot e^{-\left(\frac{\theta - \lambda_4}{\sqrt{2}\lambda_2}\right)^2} + \lambda_3 \tag{7.1}$$

where T_r is the thickness of the rust layer at coordinate θ and λ_1, λ_2, λ_3, and λ_4 are fitting parameters. The values of λ_1, λ_2, λ_3, and λ_4 from the regression analysis are presented in Table 7.1. The regression line for each sample and its corresponding 95% confidence interval are also plotted in Fig. 7.3. The fitting parameters, which are discussed in the next section, describe various characteristics of the fitted curve.

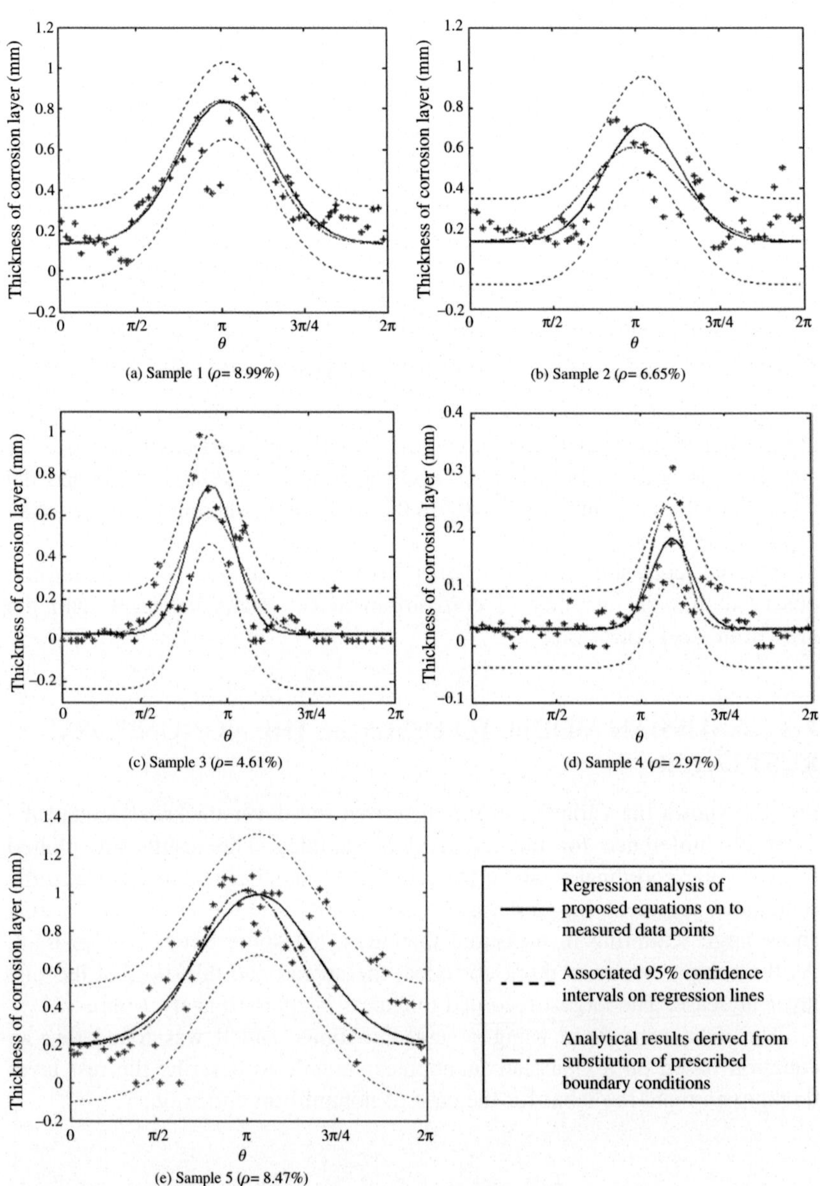

FIGURE 7.3 Measured thickness of the rust layer around the rebar perimeter.

Eq. (7.1) can be used to describe two general corrosion scenarios, as shown in Table 7.2. In the first scenario, the rust layer covers only a portion of the rebar circumference (hence, $\lambda_3 \sim 0$) as sample 3 and sample 4. The second scenario describes more severe corrosion, where the rust layer is fully

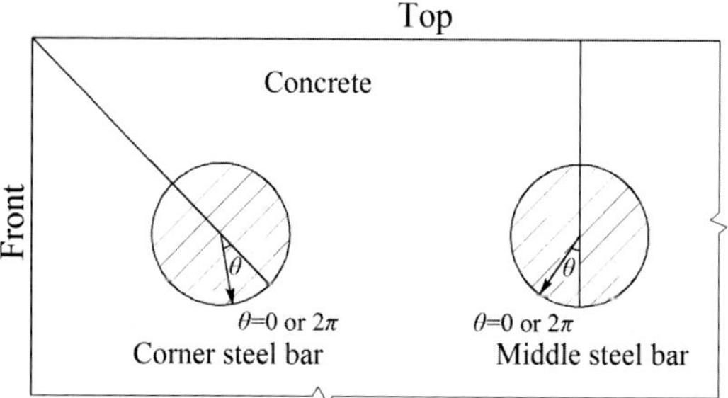

FIGURE 7.4 Polar coordinate system defined for the corner and middle rebars.

TABLE 7.1 Values of λ_1, λ_2, λ_3, and λ_4 Obtained from the Fitting of Experimental Data

Sample	Steel Corrosion, %	Regression			
		λ_1	λ_2	λ_3	λ_4
1	8.99	1.583	0.895	0.135	3.224
2	6.65	1.069	0.731	0.135	3.287
3	4.61	0.796	0.443	0.03	2.853
4	2.97	0.151	0.382	0.03	3.726
5	8.47	2.044	1.032	0.20	3.326

developed across the entire circumference of the rebar, as samples 1, 2, and 5. The parameters λ_1, λ_2, and λ_4 can also be determined analytically by substituting the boundary conditions given in Table 7.2 into Eqs. (7.2) and (7.3) for the governing corrosion scenario.

The variation in the measured thickness of the rust layer around the circumference of the corroded steel bars is described in terms of polar coordinate θ. It can be seen from Table 7.1 that the values of λ_4 vary from 2.8 to 3.7, which could be considered as approximately π. To focus on the distribution of the rust layer, the maximum thickness of the rust layer is retained at a fixed position (ie, $\theta = \pi$), as shown in Fig. 7.5a. As a result, if the

TABLE 7.2 Description of Two Nonuniform Corrosion Scenarios

	Scenario 1	Scenario 2
Proposed analytical model	$$T_r = \frac{\lambda_1}{\lambda_2 \cdot \sqrt{2\pi}} \cdot e^{-\left(\frac{\theta-\lambda_4}{\sqrt{2}\lambda_2}\right)^2} \quad (7.2)$$	$$T_r = \frac{\lambda_1}{\lambda_2 \cdot \sqrt{2\pi}} \cdot e^{-\left(\frac{\theta-\lambda_4}{\sqrt{2}a_2}\right)^2} + \lambda_3 \quad (7.3)$$
	where T_r is the thickness of the rust layer, θ is the polar coordinate, λ_3 is the minimum rust layer thickness, and λ_1, λ_2, and λ_3 are fitting coefficients.	
Boundary conditions	$T_r \cong 0, \quad \theta = \theta_1;$ $T_r \cong 0, \quad \theta = \theta_2;$ $\int_{\theta_1}^{\theta_2} T_r \mathrm{d}(\theta \cdot R) = A_r$	$T_r \cong T_0, \quad \theta = 0;$ $T_r \cong T_0, \quad \theta = 2\pi;$ $\int_0^{2\pi} T_r \mathrm{d}(\theta \cdot R) = A_r$
	where R is the radius of rebar and A_r is the total area of the rust layer.	

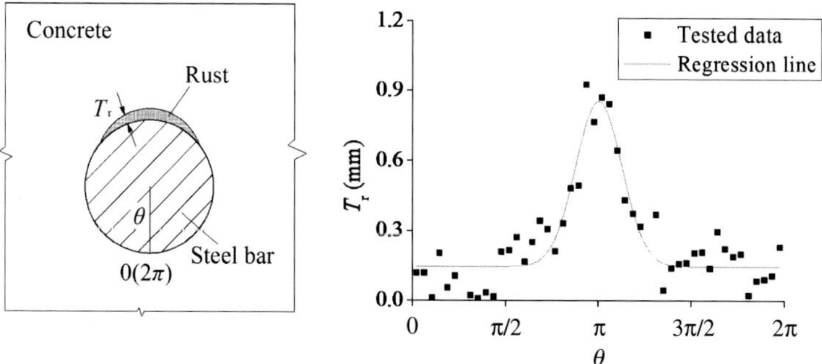

FIGURE 7.5 Polar coordinate system for measurement and fitting of the rust layer.

parameter λ_4 in the Gaussian model equals π, then the thickness of the rust layer, T_r, around the steel bar can be described as:

$$T_r = \frac{\lambda_1}{\lambda_2\sqrt{2\pi}} e^{-\left(\frac{\theta-\pi}{\sqrt{2}\lambda_2}\right)^2} + \lambda_3 \tag{7.4}$$

where T_r is the thickness of the rust layer at coordinate θ. The parameters λ_1, λ_2, and λ_3 are fitting parameters and can describe various characteristics of the rust layer. λ_1 is the nonuniform coefficient of the rust layer, λ_2 is the spread coefficient of rust layer, and λ_3 is the uniform coefficient of the rust layer.

Using slice R-5 as an example, Fig. 7.5b shows the typical variation in the measured rust layer thickness around the rebar circumference and its regression line against the polar coordinate, retaining the maximum thickness of rust layer at $\theta = \pi$. The values of parameters λ_1, λ_2, and λ_3 from the regression analysis for all samples are presented in Table 7.3. The values of steel corrosion, ρ, for all samples are also listed in Table 7.3. R^2 in Table 7.3 is the determination coefficient that can judge the fitting curve synthetically.

7.4 COMPARING THE PROPOSED GAUSSIAN MODEL WITH OTHER MODELS

The half-ellipse distribution [24,25] and the linearly decreased distribution [26] were proposed by other researchers to describe the distribution of the rust layer before concrete surface cracking. Using the tested data from the samples where the corrosion-induced cracks cannot propagate to the concrete cover surface, these three models are compared with the Gaussian model proposed by the authors.

Fig. 7.6 shows the tested data and their Gaussian, half-ellipse, and linearly decreased fitting lines. Four samples are from this experiment: R-14, R-15, M-15, and M-16; the other three are from another source [24]. The determination coefficients, R^2, for all models are shown in Fig. 7.7. The averages of R^2 are 0.826, 0.554, 0.680, and 0.638 for the Gaussian, the ellipse [24,25], and the linear [26] models, respectively. It is clear that the Gaussian model can describe nonuniform distribution of the rust layer better than the others.

TABLE 7.3 Parametric Regression Value in Model

Slice	λ_1	λ_2	λ_3	ρ, %	R^2
R-1	0.42199	0.64766	0.35231	6.07	0.35118
R-2	0.64664	0.44409	0.40586	7.07	0.63382
R-3	0.5602	0.29933	0.35546	5.89	0.81803
R-4	0.91626	0.4256	0.20282	4.71	0.8433
R-5	0.69762	0.39243	0.14528	3.50	0.85322
R-6	0.68286	1.21509	−0.01229	1.16	0.6341
R-7	0.15277	0.47834	0.01139	0.55	0.44181
R-8	0.35013	0.38399	0.00424	0.75	0.9308
R-9	0.19531	0.17344	0.01329	0.71	0.85532
R-10	1.26435	0.81353	−0.01504	2.34	0.8825
R-11	1.43298	1.10063	−0.03442	2.78	0.85479
R-12	0.99129	0.75861	−0.00787	2.20	0.90918
R-13	0.29946	0.42558	−0.00052	0.64	0.92299
R-14	0.73464	1.65737	−0.03762	0.53	0.77982
R-15	0.3171	0.54062	0.00716	0.76	0.908
R-16	0.86091	1.78469	−0.05184	1.06	0.77792
R-17	0.32806	1.04671	0.02558	1.13	0.615
R-18	0.46704	0.72154	0.05279	1.96	0.81462
R-19	0.37664	0.81054	0.05261	1.63	0.75833
R-20	1.48078	2.00202	−0.03037	2.61	0.70365
M-15	0.56919	0.53812	0.00479	1.38	0.81723
M-16	1.22506	1.15475	−0.04912	2.16	0.67082

7.5 PARAMETERS IN GAUSSIAN MODEL

7.5.1 λ_3: Uniform Coefficient of the Rust Layer

When a steel bar has uniform rust, the thickness of the rust layer, T_r, should be a constant value. In Eq. (7.4), $\frac{\lambda_1}{\lambda_2\sqrt{2\pi}}e^{-(\theta-\pi/\sqrt{2}\lambda_2)^2}$ is the part related to coordinate θ; under this uniform rust circumstance, if $\frac{\lambda_1}{\lambda_2\sqrt{2\pi}}e^{-(\theta-\pi/\sqrt{2}\lambda_2)^2}=0$, then $T_r=\lambda_3$. λ_3, defined as the uniform coefficient, is independent of the value of θ and can describe the thickness of the rust layer, which covers the entire circumference of the steel bar.

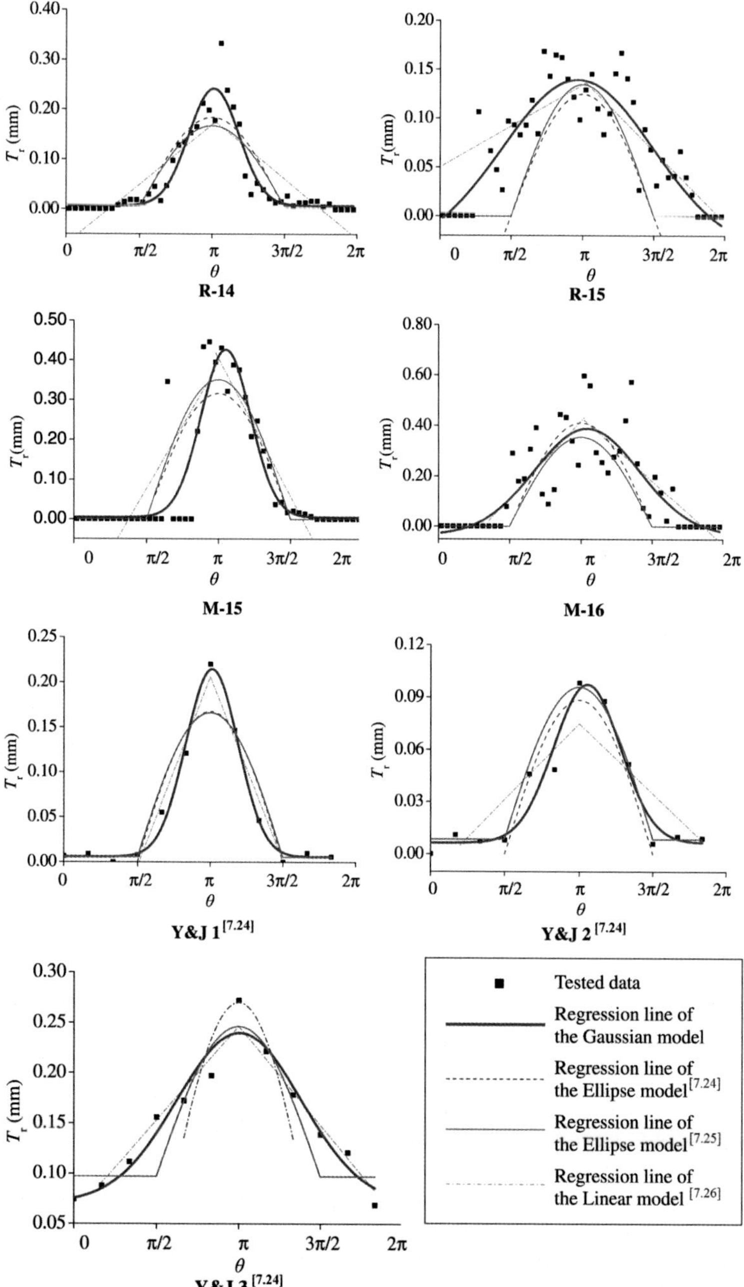

FIGURE 7.6 The regression analysis of the proposed models for the tested data.

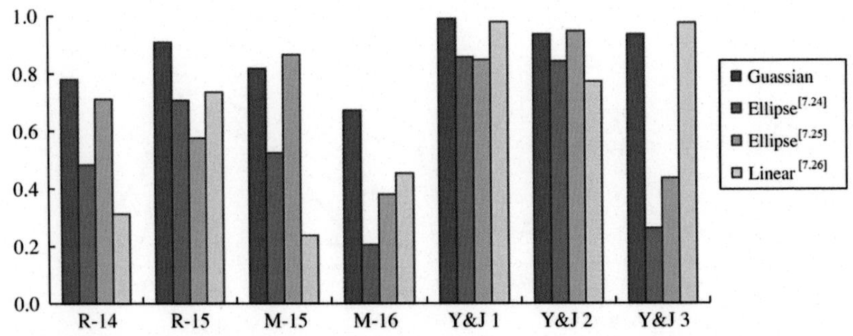

FIGURE 7.7 R^2 of four models.

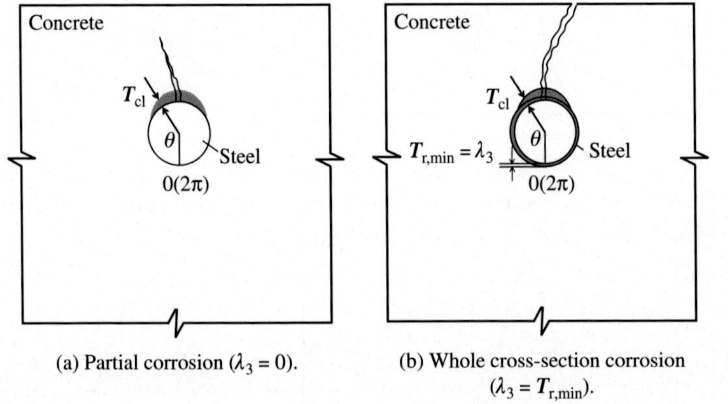

FIGURE 7.8 Physical meaning of λ_3. (a) Partial corrosion ($\lambda_3 = 0$). (b) Whole cross-section corrosion ($\lambda_3 = T_{r,min}$).

For corrosion products that cover only a portion of the rebar circumference, including the samples R-6, R-7, R-8, R-9, R-10, R-11, R-12, R-13, R-14, R-15, R-16, M-15, and M-16, the values of λ_3 are close to 0, as can be seen from Table 7.3. Therefore, when the steel bar is in partial corrosion, the corrosion layer λ_3 is taken as 0 (Fig. 7.8a). When the corrosion products develop across the entire circumference of the rebar, including the samples R-1, R-2, R-3, R-4, R-5, R-17, R-18, R-19, and R-20, λ_3 (in mm) represents the minimum thickness of the corrosion layer, as illustrated in Fig. 7.8b.

7.5.2 λ_1: Nonuniform Coefficient of the Rust Layer

The parameter λ_3 reflects the uniform rust level, as discussed in Section 7.5.1. $\lambda_3 = 0$ when the steel bar has partial rust under the partial rust circumstance:

$$T_r = \frac{\lambda_1}{\lambda_2\sqrt{2\pi}} e^{-\left(\frac{\theta-\pi}{\sqrt{2}\lambda_2}\right)^2} \tag{7.5}$$

To calculate the area enclosed by the function, θ is integrated from 0 to 2π:

$$\begin{aligned} \int_0^{2\pi} T_r \mathrm{d}\theta &= \int_0^{2\pi} \frac{\lambda_1}{\lambda_2\sqrt{2\pi}} e^{-\left(\frac{\theta-\pi}{\sqrt{2}\lambda_2}\right)^2} \mathrm{d}\theta \\ &= \int_0^{2\pi} \frac{\lambda_1}{\sqrt{\pi}} e^{-\left(\frac{\theta-\pi}{\sqrt{2}\lambda_2}\right)^2} \mathrm{d}\left(\frac{\theta-\pi}{\sqrt{2}\lambda_2}\right) \\ &= \lambda_1 \times \mathrm{erf}\left(\frac{\pi}{\sqrt{2}\lambda_2}\right) \end{aligned} \tag{7.6}$$

where the error function is an increasing function from -1 to 1. Its value gets closer to 1 when the value of λ_2 becomes smaller. The values of λ_2 for most samples are smaller than 1.5, as shown in Table 7.3. Taking λ_2 as 1.5, $\mathrm{erf}(\pi/1.5\times\sqrt{2}) = 0.964$; therefore, for most samples, the values of $\mathrm{erf}(\pi/\sqrt{2}\lambda_2)$ are larger than 0.964, so that $\int_0^{2\pi} T_r \mathrm{d}\theta$ is very close to λ_1, indicating that λ_1 is approximately the area of the peak of the Gaussian function. Hence, λ_1 is defined as the nonuniform coefficient, reflecting the nonuniform rust level.

Two scenarios of steel corrosion are discussed in the following sections.

7.5.2.1 Corrosion Develops Across the Entire Steel Circumference

The distribution of the rust layer can be subdivided by two parts, as illustrated in Fig. 7.9. The bottom part of Fig. 7.9a shows the corrosion products uniformly distributed across the whole perimeter of the steel bar, which can be described by λ_3. The top peak part of Fig. 7.9 is related to λ_1. With the increase of λ_1, the area of corrosion peaks grows, as shown in Fig. 7.10, meaning that the steel corrosion accommodates more locally.

7.5.2.2 Corrosion Covers a Portion of the Steel Circumference

In this scenario, the corrosion products cover only a part of the perimeter of the rebar, as shown in Fig. 7.11. Fig. 7.12 shows that λ_1 is linearly proportional to the steel corrosion ρ as follows:

$$\lambda_1 = 49\rho \tag{7.7}$$

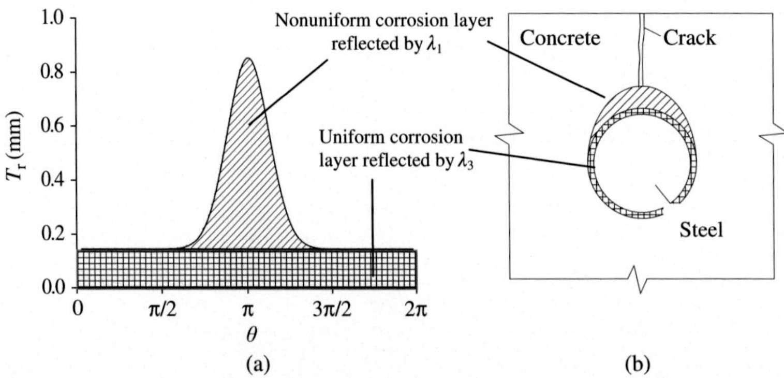

FIGURE 7.9 Two parts of the rust layer when steel corrosion spreads throughout the entire circumference.

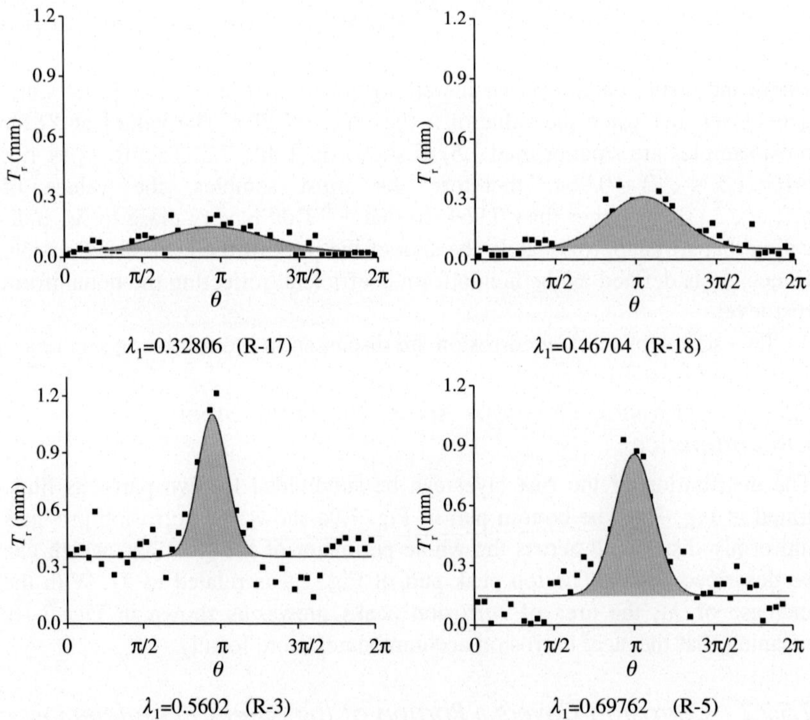

FIGURE 7.10 Area of corrosion peaks grow with the increase of λ_1.

7.5.3 λ_2: Spread Coefficient of Rust Layer

The parameter λ_2 describes the spreading range of nonuniform corrosion. Fig. 7.13 shows that as λ_2 increases, the rust spreads more extensively,

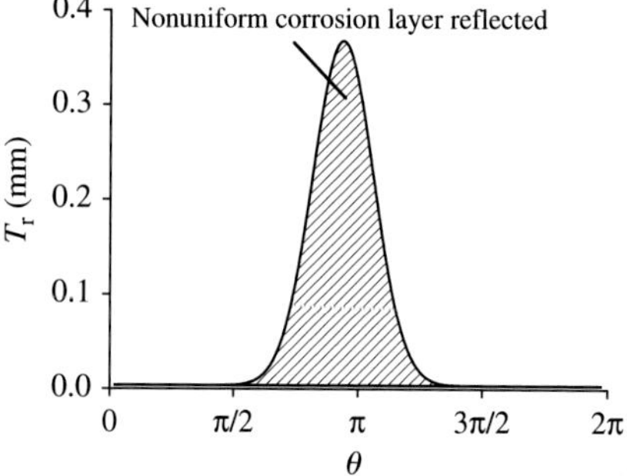

FIGURE 7.11 Peak area of partial corrosion.

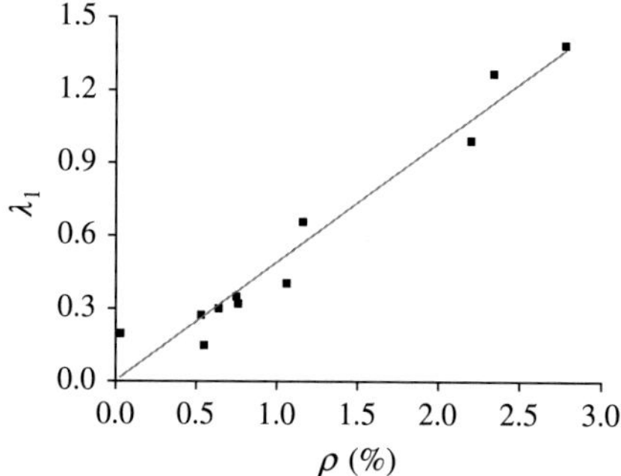

FIGURE 7.12 Relationship between λ_1 and ρ.

leading to a flatter shape of the rust layer. However, the small value of this parameter, for example, sample R-9, results in a steep shape of the rust layer. This means that the peak value of the rust layer, $T_{r,max}$, could be jointly determined by λ_1 and λ_2. For instance, when λ_1 is the same, smaller λ_2 results in larger $T_{r,max}$. The relationship among λ_1, λ_2, and $T_{r,max}$ is discussed in detail in Section 7.5.4.2.

7.5.4 Relationships Among Parameters Before Concrete Surface Cracking

7.5.4.1 Relationship Between λ_1 and ρ

Before cracks penetrate the concrete cover, corrosion products cover only a portion of the circumference of the steel bar and λ_3 equals 0. Under this circumstance, Eq. (7.4) can be simplified as:

$$T_r = \frac{\lambda_1}{\lambda_2\sqrt{2\pi}} e^{-\left(\frac{\theta-\pi}{\sqrt{2}\lambda_2}\right)^2} \tag{7.8}$$

The area of rust layer can be obtained by integrating Eq. (7.8)

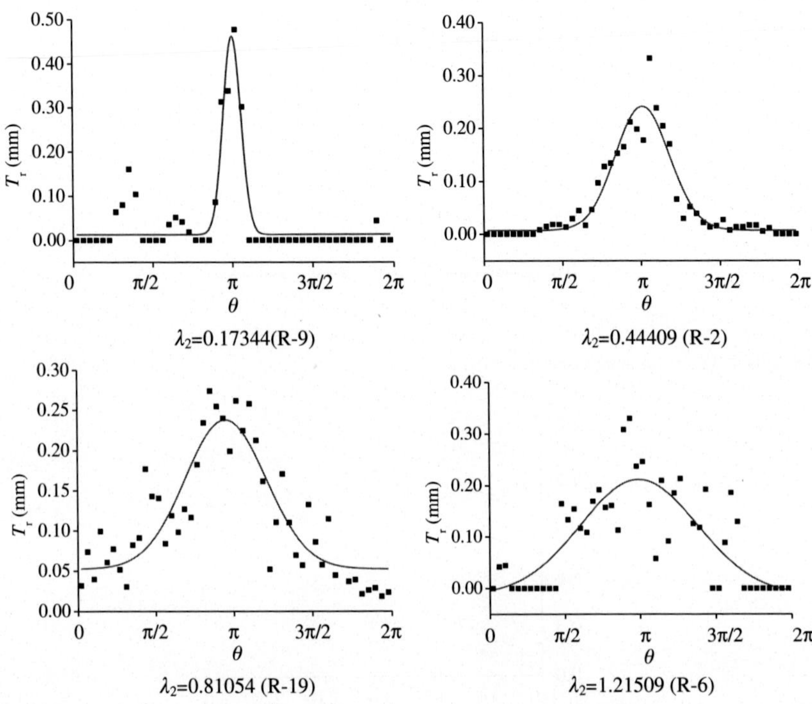

FIGURE 7.13 Nonuniform corrosion spreading widely with the increase of λ_2.

$$A_r = n \cdot A_s \cdot \rho = \int_{-\infty}^{+\infty} T_r R d\theta \tag{7.9}$$

where A_r is the total area of rust layer, A_s is the cross-section area of uncorroded steel bar, n is the volume expansion ratio of rust and the original steel, and ρ is steel corrosion.

Substituting Eq. (7.8) into Eq. (7.9), the following equation can be obtained:

$$\rho = \frac{\lambda_1 \cdot R}{\sqrt{\pi} n \cdot A_s} \int_{-\infty}^{+\infty} e^{-t^2} dt \tag{7.10}$$

The right side of Eq. (7.10) is the integration of a transcendental function:

$$\int_{-\infty}^{+\infty} e^{-t^2} dt = \sqrt{\pi} \tag{7.11}$$

Eq. (7.10) can then be simplified as:

$$\rho = \frac{\lambda_1 \cdot R}{n \cdot A_s} \tag{7.12}$$

Hence, λ_1 can be expressed as:

$$\lambda_1 = \frac{\rho \cdot R}{n \cdot A_s} \tag{7.13}$$

For a reinforced concrete component, its n and A_s are constants. Therefore, Eq. (7.13) reflects that λ_1 is linearly proportional to the steel corrosion ρ.

7.5.4.2 Relationship Among λ_1, λ_2, and $T_{r,max}$

The maximum thickness of the nonuniform rust layer, $T_{r,max}$ (ie, the peak value of the rust layer) has a great influence on the development of concrete cover cracking. The larger $T_{r,max}$ leads to quick cracking of the concrete cover.

To analyze the relationship between λ_1 and λ_2, let $\lambda_3 = 0$ in the following analysis. According to Eq. (7.4), the thickness of rust layer T_r reaches its maximum value at $\theta = \pi$:

$$T_{r,max} = \frac{\lambda_1}{\lambda_2\sqrt{2\pi}} \tag{7.14}$$

Eq. (7.14) reveals that there is a linear relationship between λ_1/λ_2 and $T_{r,max}$. It also explains the finding in Section 7.5.3 that smaller λ_2 results in larger $T_{r,max}$ when λ_1 is same.

7.5.4.3 Relationship Among λ_1, λ_3, and T_r

The Gaussian Eq. (7.4) is integrated to obtain the total area of the rust layer:

$$\begin{aligned} A_r &= \int_0^{2\pi} T_r \cdot R d\theta = \int_0^{2\pi} \left(\frac{\lambda_1}{\lambda_2\sqrt{2\pi}} e^{-\left(\frac{\theta-\pi}{\sqrt{2}\lambda_2}\right)^2} + \lambda_3 \right) R d\theta \\ &= \int_0^{2\pi} \left(\frac{\lambda_1}{\lambda_2\sqrt{2\pi}} e^{-\left(\frac{\theta-\pi}{\sqrt{2}\lambda_2}\right)^2} \right) R d\theta + \int_0^{2\pi} \lambda_3 R d\theta \\ &= \int_0^{2\pi} \left(\frac{\lambda_1}{\sqrt{\pi}} e^{-\left(\frac{\theta-\pi}{\sqrt{2}\lambda_2}\right)^2} \right) R d\left(\frac{\theta-\pi}{\sqrt{2}\lambda_2} \right) + \int_0^{2\pi} \lambda_3 R d\theta \\ &= \lambda_1 R \times \operatorname{erf}\left(\frac{\pi}{\sqrt{2}\lambda_2} \right) + 2\pi\lambda_3 R \\ &\approx (\lambda_1 + 2\pi\lambda_3) R \end{aligned} \tag{7.15}$$

where A_r is the area of the rust layer (mm^2) and R is the radius of the steel bar (mm).

According to the geometric relationship, the thickness of the rust layer is:

$$\overline{T_r} = \frac{A_r}{2\pi R} = \frac{(\lambda_1 + 2\pi\lambda_3)R}{2\pi R} = \frac{1}{2\pi}(\lambda_1 + 2\pi\lambda_3) \quad (7.16)$$

It can be seen from Eq. (7.16) that the thickness of the rust layer shows a linear relationship with $(\lambda_1 + 2\pi\lambda_3)$.

7.6 CONCLUSIONS

1. The Gaussian model has a better ability to describe the variability in the thickness of the rust layer deposited on the circumference of a steel bar. The parameters in the Gaussian model (ie, the nonuniform coefficient λ_1, the spread coefficient λ_2, and the uniform coefficient λ_3) can describe the nonuniform corrosion level, the spreading range of nonuniform corrosion, and the uniform corrosion level of the rust layer deposited around the perimeter of rebar.
2. The analysis and discussion of parameters in the Gaussian model reveal the following:
   - the nonuniform coefficient λ_1 is linearly proportional to the steel rust ρ;
   - the uniform coefficient λ_3 has a linear relationship with the minimum thickness of the rust layer $T_{r,min}$;
   - λ_1/λ_2 shows a linear relationship with the maximum thickness of the rust layer $T_{r,max}$;
   - the thickness of the rust layer T_r has a linear relationship with $(\lambda_1 + 2\pi\lambda_3)$.

REFERENCES

[1] Bazant ZP. Physical model for steel corrosion in concrete sea structures—application. J Struct Div 1979;105 (ASCE 14652 Proceeding).

[2] Liu Y, Weyers RE. Modeling the time-to-corrosion cracking in chloride contaminated reinforced concrete structures. ACI Mater J 1998;95(6):675–81.

[3] Pantazopoulou SJ, Papoulia KD. Modeling cover-cracking due to reinforcement corrosion in RC structures. J Eng Mech 2001;127(4):342–51.

[4] Bhargava K, Ghosh AK, Mori Y, et al. Modeling of time to corrosion-induced cover cracking in reinforced concrete structures. Cement Concrete Res 2005;35(11):2203–18.

[5] Zhao YX, Jin WL. Modeling the amount of steel corrosion at the cracking of concrete cover. Adv Struct Eng 2006;9(5):687–96.

[6] Bhargava K, Ghosh AK, Mori Y, et al. Model for cover cracking due to rebar corrosion in RC structures. Eng Struct 2006;28(8):1093–109.

[7] Li C, Melchers RE, Zheng J. Analytical model for corrosion-induced crack width in reinforced concrete structures. ACI Struct J 2006;103(4):479–87.

[8] El Maaddawy T, Soudki K. A model for prediction of time from corrosion initiation to corrosion cracking. Cement Concrete Composites 2007;29(3):168–75.
[9] Chernin L, Val DV, Volokh KY. Analytical modelling of concrete cover cracking caused by corrosion of reinforcement. Mater Struct 2010;43(4):543–56.
[10] Kim KH, Jang SY, Jang BS, et al. Modeling mechanical behavior of reinforced concrete due to corrosion of steel bar. ACI Mater J 2010;107(2).
[11] Lu C, Jin W, Liu R. Reinforcement corrosion-induced cover cracking and its time prediction for reinforced concrete structures. Corrosion Sci 2011;53(4):1337–47.
[12] Rasheeduzzafar, Al-Saadoun SS, Al-Gahtani AS. Corrosion cracking in relation to bar diameter, cover, and concrete quality. J Mater Civil Eng 1992;4(4):327–42.
[13] Andrade C, Alonso C, Molina FJ. Cover cracking as a function of bar corrosion: Part I-experimental test. Mater Struct 1993;26(8):453–64.
[14] Alonso C, Andrade C, Rodriguez J, et al. Factors controlling cracking of concrete affected by reinforcement corrosion. Mater Struct 1998;31(7):435–41.
[15] Zhang WP. Damage prediction and durability estimation for corrosion of reinforcement in concrete structures. Shanghai: Tongji University; 1999 [in Chinese].
[16] Vidal T, Castel A, Francois R. Analyzing crack width to predict corrosion in reinforced concrete. Cement Concrete Res 2004;34(1):165–74.
[17] Zhang R, Castel A, François R. Concrete cover cracking with reinforcement corrosion of RC beam during chloride-induced corrosion process. Cement Concrete Res 2010;40(3):415–25.
[18] Asami K, Kikuchi M. In-depth distribution of rusts on a plain carbon steel and weathering steels exposed to coastal–industrial atmosphere for 17 years. Corrosion Sci 2003;45(11):2671–88.
[19] Duffó GS, Morris W, Raspini I, et al. A study of steel rebars embedded in concrete during 65 years. Corrosion Sci 2004;46(9):2143–57.
[20] Chitty W, Dillmann P, L'Hostis V, et al. Long-term corrosion resistance of metallic reinforcements in concrete—a study of corrosion mechanisms based on archaeological artefacts. Corrosion Sci 2005;47(6):1555–81.
[21] Zhao YX, Ren HY, Dai H, et al. Composition and expansion coefficient of rust based on X-ray diffraction and thermal analysis. Corrosion Sci 2011;53(5):1646–58.
[22] Wong HS, Zhao YX, Karimi AR, et al. On the penetration of corrosion products from reinforcing steel into concrete due to chloride-induced corrosion. Corrosion Sci 2010;52(7):2469–80.
[23] Zhao YX, Karimi AR, Wong HS, et al. Comparison of uniform and non-uniform corrosion induced damage in reinforced concrete based on a Gaussian description of the corrosion layer. Corrosion Sci 2011;53(9):2803–14.
[24] Yuan YS, Ji YS, Mu YJ. Propagation and model of distribution for corrosion of steel bars in concrete. China Civil Eng J 2007;40(7):5–10.
[25] Malumbela G, Alexander M, Moyo P. Model for cover cracking of RC beams due to partial surface steel corrosion. Constr Building Mater 2011;25(2):987–91.
[26] Jang BS, Oh BH. Effects of non-uniform corrosion on the cracking and service life of reinforced concrete structures. Cement Concrete Res 2010;40(9):1441–50.

Chapter 8

Crack Shape of Corrosion-Induced Cracking in the Concrete Cover

Chapter Outline

8.1 INTRODUCTION

The analytical and numerical models cannot accurately reflect the corrosion-induced cracking process without considering the crack patterns and shapes in the concrete cover induced by steel corrosion. In this chapter, the shape of the corrosion-induced cracks in the concrete cover is investigated. Four groups of reinforced concrete specimens, including natural aggregate concrete (NAC) and recycled aggregate concrete (RAC), with different replacement percentages of recycled aggregate (RA) were cast, and the corrosion-induced crack development in each concrete cover was investigated.

Steel Corrosion-Induced Concrete Cracking. DOI: http://dx.doi.org/10.1016/B978-0-12-809197-5.00008-6

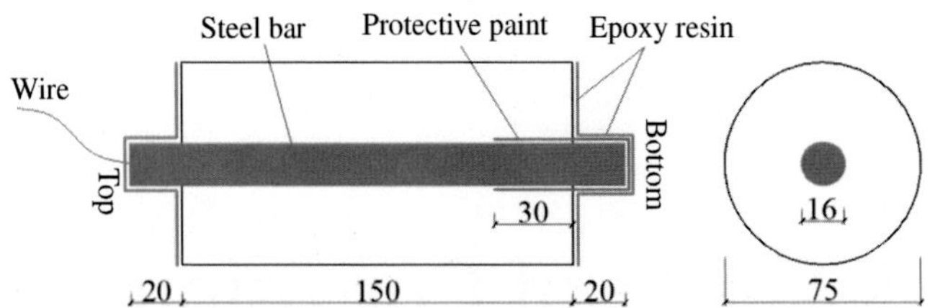

FIGURE 8.1 Layout details of specimens (dimensions are in mm).

TABLE 8.1 Compositions of the Concrete Specimen Mixtures

Specimens	Component (kg/m^3)					f_c (MPa)
	Cement	Water	Sand	NA	RA	
R000	430	185	559	1118	0	30.4
R033	430	190	559	745	373	28.9
R067	430	195	559	373	745	27.4
R100	430	200	559	0	1118	26.2

8.2 EXPERIMENTAL PROGRAM

8.2.1 Reinforced Concrete Specimens

Cylindrical reinforced concrete specimens with 0%, 33%, 67%, and 100% replacement of natural aggregate (NA) with RA were used in the study; the dimensions are shown in Fig. 8.1. Each specimen had a diameter of 75 mm and a height of 150 mm, with a 16-mm-diameter plain hot-rolled steel bar in the center. To avoid the corrosion of steel at the end of each specimen, both ends of the specimens and the exposed steel bars were coated with epoxy resin. The top end of each steel bar was attached to a wire, and the bottom end was treated with protective paint to prevent corrosion products from dissolving in the NaCl solution in which the specimens were to be submerged. Two specimens were cast at each percentage of RA replacement and were labeled RXXX-N, where XXX was the percentage of RA replacement and N was the specimen number for a given RA replacement percentage. For example, R067-1 represents the first specimen with an NA replacement rate of 67%.

The mixture proportions of the four groups of concrete are shown in Table 8.1. The binder was ordinary Portland cement with a grade of 42.5R. The fine aggregate was medium natural river sand with a fineness modulus of 2.5. The NA was normal gravel, and the RA was a commercial product purchased from Shanghai, China. Both the NA and RA consisted of coarse particles with a size range of 5–25 mm. Due to the greater water absorption of RA, extra water, calculated by multiplying the water needed for the NA mixtures by 1.34%, was added to the mixtures to reach the target slump for the fresh concrete, as shown in Table 8.1. After casting, all of the specimens were covered with damp burlap and wetted once per day for 28 days. Table 8.1 also shows the 28-day compressive strength (f_c) values of the concrete mixtures.

8.2.2 Accelerated Corrosion History

After 28 days of curing, the specimens were wrapped in stainless steel nets with sponge material filling the spaces between the specimens and the nets. The wetting and drying cycles were coupled with DC power to accelerate the steel corrosion. The specimens were subjected to a 1-day wetting period followed by a 2-day drying period for each cycle. During the wetting period, the specimens were partially immersed in a 3.5% NaCl solution so that the sponge would remain wet (as shown in Fig. 8.2), and a 300 $\mu A/cm^2$ current was applied between the reinforcing steel bars (acting as anodes) in the specimens and the stainless steel nets (acting as cathodes). During the drying periods, the DC power was turned off and the specimens were removed from the sponge and exposed to air in the laboratory for 2 days.

After five cycles, corrosion-induced cracks running approximately parallel to the center bar inside the specimens were observed. The DC power source was then turned off and the specimens were removed for further investigation. It should be noted that the concrete cover of specimen R100-2 was totally damaged by steel corrosion and could not be used for further investigation.

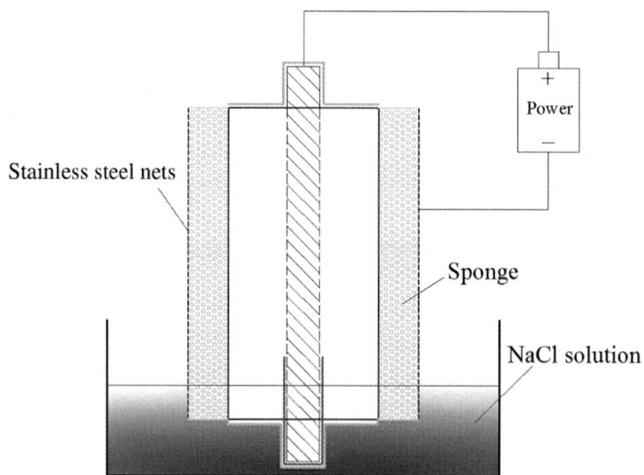

FIGURE 8.2 Wetting and drying cycles combined with a constant current.

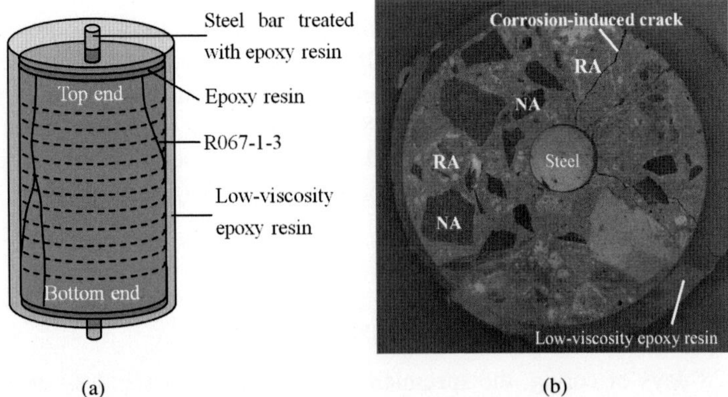

FIGURE 8.3 Schematic diagram of preparation of samples for digital microscopy observation. (a) Cracked parts of specimens were cast in epoxy resin. (b) A sample prepared for digital microscopy observation.

8.2.3 Sample Preparation

The cracked portions of the specimens were cast in a low-viscosity epoxy resin to minimize any artificial damage that might occur. Each specimen was then carefully cut into 10-mm-thick slices, as shown in Fig. 8.3a, from top to bottom (except for the two end parts) with a precision saw (SYJ-200). One of those slices is shown in Fig. 8.3b. The slices were labeled as RXXX-N-M. For example, R067-1-3 represents the third slice from the specimen R067-1, as shown in Fig. 8.3a. Approximately 10 slices were obtained from each specimen; however, some were damaged during the cutting process and only four to six slices remained from each specimen. The slices were polished with a precision grinding and polishing machine (UNIPOL-1502). To prevent further corrosion, all samples were kept in a dry environment (relative humidity less than 30%) for several days before observation.

8.2.4 Observation and Measurement

8.2.4.1 *Crack Width*

The information concerning concrete cracking primarily relied on the visual observation of cracks, so microscopic cracking was neglected. The widths of visible cracks were measured with digital microscopy. Measurements were made for each slice along the circumferential direction at intervals of 1 mm in the radial direction from the surface of the steel bar toward the concrete surface, as shown in Fig. 8.4a. The total crack width, W_i (in mm), the sum of

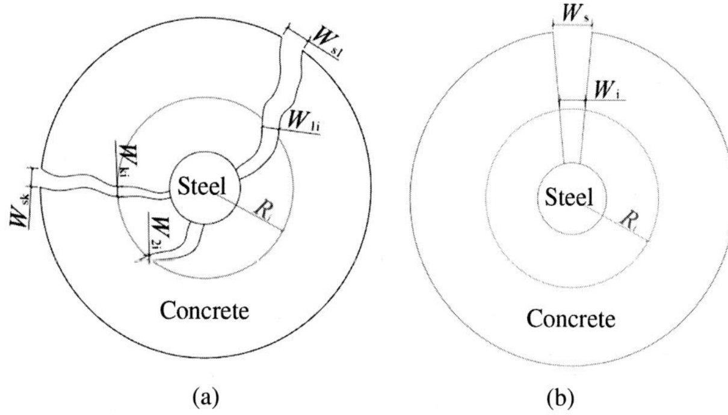

FIGURE 8.4 Measurement and calculation of the crack width, W_i, at radius R_i. (a) Measurement of crack width. (b) Total crack width, W_i.

the entire crack widths at radius, R_i (in mm), as shown in Fig. 8.4b, can be expressed as follows:

$$W_i = \sum_{k=1}^{j} w_{ki} \tag{8.1}$$

where W_i is the total circumferential crack width of radius R_i (in mm), w_{ki} is the circumferential crack width of a single crack at radius R_i (in mm), and j is the total number of cracks at radius R_i. The relationship between the total circumferential crack width and the radius can be expressed as:

$$W_i = f(R_i) \tag{8.2}$$

In addition, the visible crack widths on the surface of the concrete cover were also measured. For each slice, the total crack width, W_s (in mm), is the sum of all of the concrete surface crack widths, as shown in Fig. 8.4, and can be expressed as follows:

$$W_s = \sum_{k=1}^{j} w_{sk} \tag{8.3}$$

where w_{sk} is the crack width of a single crack at the edge of a slice (in mm) and j is the total number of cracks on the surface of the concrete cover.

8.2.4.2 Corrosion Layer Thickness

The area of the rust layer, A_r, was measured with a digital microscope (Promicro scan 5866) because the rust layer can be clearly distinguished in color images with a digital microscope connected to a computer. It should be noted that the mill scale did not contribute to the volume expansion acting on the surrounding concrete cover as stated in chapter "Mill Scale and Corrosion Layer at Concrete Surface Cracking." Therefore, the thickness of the mill scale, T_m, should be subtracted from the thickness of rust layer, T_r, when calculating the thickness of the corrosion layer, T_{CL}. The mean thickness of the mill scale of this batch of steel bar was 0.0275 mm. The mean thickness of the corrosion layer, T_{CL}, can then be calculated as follows:

$$T_{CL} = T_r - T_m = \frac{A_r}{2\pi R} - T_m \tag{8.4}$$

8.3 CRACK SHAPE

8.3.1 Crack Width Model

Using slices R000-1-8 and R000-2-8 as examples, Fig. 8.5 shows the typical variation in the measured circumferential crack width along the direction of the radius and the regression line of each slice. In slice R000-1-8, the corrosion-induced cracks did not propagate to the outer surface of the concrete cover, whereas the crack penetrated the concrete cover of slice R000-2-8. It can be observed that the total circumferential crack width is linearly proportional to the radius. Therefore, the linear crack shape model is adopted here as follows:

$$W_i = a_1 l_{ci} + a_2 \tag{8.5}$$

where l_{ci} is the distance from radius R_i to the surface of the steel bar (in mm), and $l_{ci} = R_i - R$; the parameters a_1 and a_2 can be physically interpreted to describe various characteristics of the cracks. The physical meaning of a_1 is the crack width variation coefficient, and that of a_2 is the crack width coefficient at the surface of the steel bar, as shown schematically in Fig. 8.6.

The values of corrosion layer thickness, T_{CL}, the parameters a_1 and a_2 obtained from the regression analysis, and the measured crack width on the surface of the concrete cover W_s of all the slices are presented in Table 8.2.

8.3.2 a_1: Crack Width Variation Coefficient

The parameter a_1 describes the variation in the crack width along the radial direction, and a_1 is linearly proportional to the corrosion layer thickness, T_{CL}, regardless of the replacement of RA in the concrete, as shown Fig. 8.7. The details of the linear regression equations of the four types of concrete are listed in Table 8.3. Table 8.3 shows that the slopes of the regression lines increase with the increase of the RA replacement percentage, which means that the use of RA results in a crack shape with a wider opening at the same steel corrosion level.

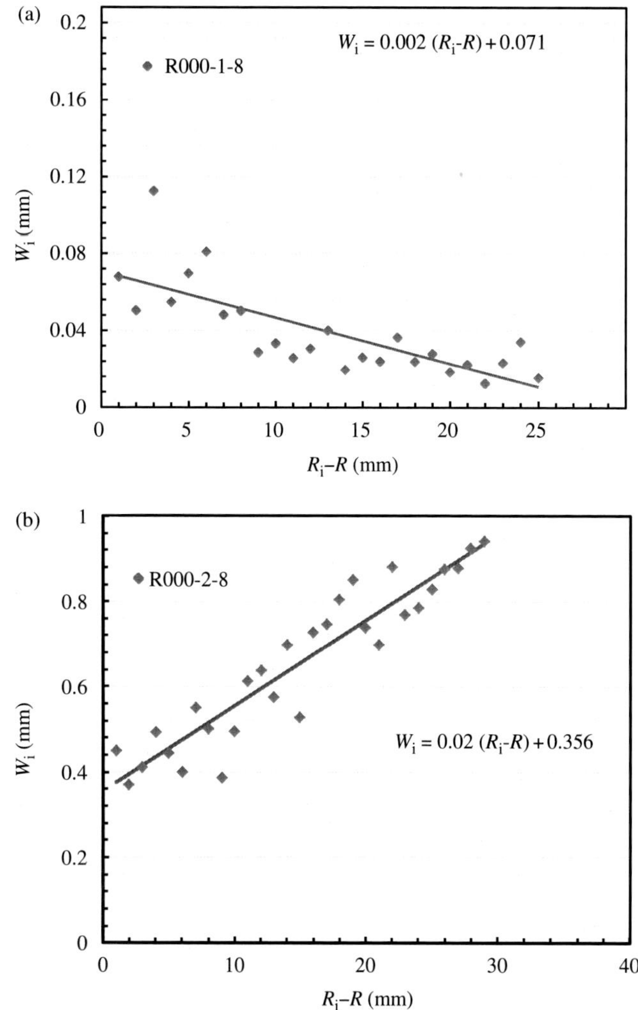

FIGURE 8.5 Measured data and the fitting line of the crack width. (a) Slice R000-1-8, representing the inner cracking scenario. (b) Slice R000-2-8, representing the cracks that had penetrated the concrete cover.

It can be seen in Table 8.2 that $a_1 < 0$ before the crack reaches the concrete outer surface (slice R000-1-8), and that $a_1 > 0$ after concrete surface cracking. Therefore, we infer that a_1 might be 0 at the moment of concrete surface cracking. At this moment, $T_{\mathrm{CL}} = T_{\mathrm{CL}}^{\mathrm{surface}}$, where $T_{\mathrm{CL}}^{\mathrm{surface}}$ is the critical rust layer thickness at concrete surface cracking. For the different groups of concrete specimens, the values of $T_{\mathrm{CL}}^{\mathrm{surface}}$, which are discussed later in Section 8.5, are also listed in Table 8.3.

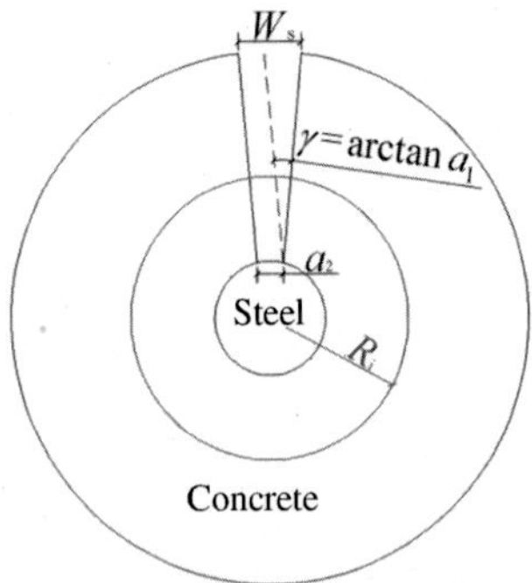

FIGURE 8.6 Schematic crack shape model.

TABLE 8.2 Measurement Results and Parametric Regression Values

Slice	T_{CL} (mm)	a_1	a_2	W_s(mm)	NOS	NIC
R000-1-4	0.0489	0.003	0.101	0.702	3	2
R000-1-5	0.3141	0.027	0.121	1.549	3	1
R000-1-6	0.1061	0.007	0.277	0.384	3	0
R000-1-8	0.0229	-0.002	0.071	0	0	1
R000-2-3	0.1539	0.008	0.325	0.541	3	1
R000-2-5	0.2954	0.026	1.578	2.035	5	1
R000-2-6	0.2267	0.007	0.525	1.013	3	1
R000-2-7	0.3120	0.028	1.267	1.822	5	2
R000-2-8	0.1560	0.02	0.356	0.811	2	2
R033-1-4	0.6022	0.053	1.075	2.916	6	4
R033-1-5	0.3484	0.043	1.295	2.662	3	1
R033-1-6	0.1841	0.007	1.076	1.304	2	1
R033-1-7	0.0406	0.002	0.139	0.151	1	0
R033-1-8	0.0416	0.002	0.141	0.164	1	1
R033-2-5	0.6261	0.059	1.834	2.703	6	1
R033-2-6	0.5866	0.046	1.964	3.924	7	1
R033-2-7	0.3505	0.027	1.495	2.508	4	2
R033-2-8	0.3983	0.032	1.024	2.685	4	0
R067-1-1	0.3234	0.033	0.932	1.613	3	2

(Continued)

TABLE 8.2 (Continued)

Slice	T_{CL} (mm)	a_1	a_2	W_s(mm)	NOS	NIC
R067-1-2	0.2049	0.013	0.871	1.077	5	1
R067-1-3	0.2132	0.019	0.604	1.352	3	3
R067-1-4	0.7582	0.076	2.436	4.444	7	2
R067-2-3	0.8570	0.089	2.122	5.953	7	1
R067-2-4	0.7093	0.056	2.265	3.936	7	2
R067-2-5	0.5242	0.047	1.019	2.895	6	2
R067-2-6	0.4930	0.044	1.93	3.011	5	2
R067-2-7	0.4805	0.042	1.524	2.565	6	2
R067-2-8	0.3141	0.039	1.205	1.819	4	1
R100-1-2	0.8403	0.095	2.666	5.186	6	1
R100-1-3	1.0161	0.09	3.329	6.956	7	1
R100-1-4	1.1575	0.092	3.274	7.942	9	2
R100-1-5	1.2948	0.11	4.018	8.133	9	4
R100-1-6	1.1492	0.145	2.981	5.962	8	2

Note: NOS stands for the number of cracks that have penetrated the concrete cover surface, and NIC stands for the number of inner cracks.

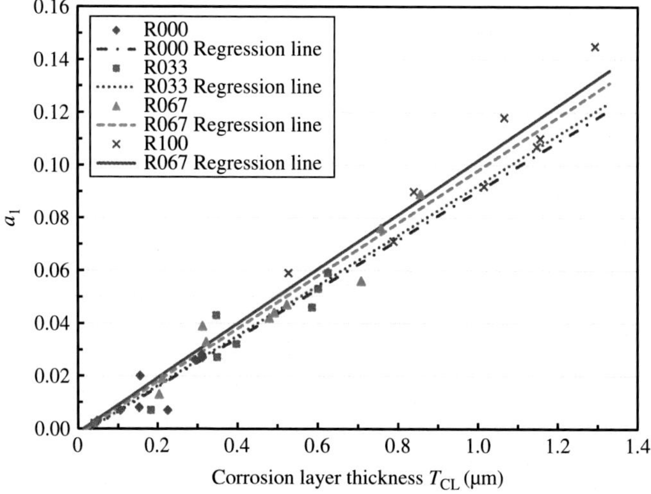

FIGURE 8.7 Relationship between parameter a_1 and corrosion layer thickness T_{CL}.

TABLE 8.3 Linear Regression Results for a_1 Compared to T_{CL}

Specimens	Linear Regression Equations	R^2	$T_{CL}^{surface}$ (mm)
R000	$a_1 = 0.0919T_{CL} - 0.0029$	0.795	0.0316
R033	$a_1 = 0.0931T_{CL} - 0.0028$	0.925	0.0301
R067	$a_1 = 0.0981T_{CL} - 0.0021$	0.927	0.0214
R100	$a_1 = 0.1025T_{CL} - 0.0014$	0.868	0.0137

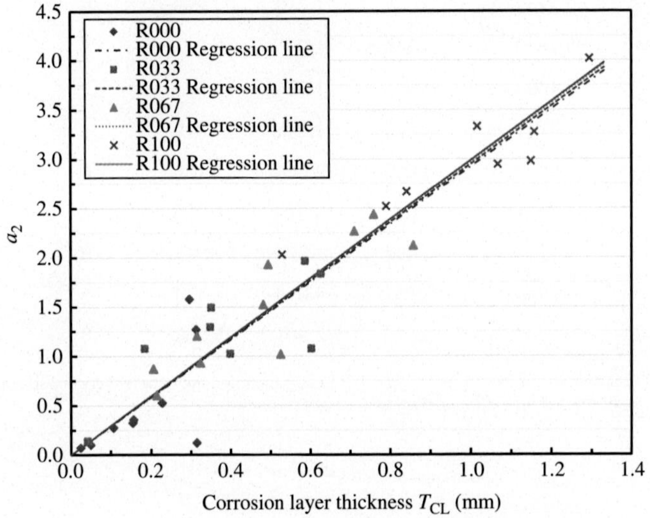

FIGURE 8.8 Relationship between parameter a_2 and corrosion layer thickness T_{CL}.

8.3.3 a_2: Crack Width Coefficient at the Surface of the Steel Bar

Parameter a_2 reflects the crack width on the surface of the steel bar, as shown in Fig. 8.6. It can be seen from Fig. 8.8 and Table 8.4 that parameter a_2 increases linearly with the increase of the steel corrosion regardless of the replacement of RA in the concrete mixture, and the differences of the fitting line of the four types of concrete are not significant, indicating that crack shapes of inner cracks for different types of concrete are mostly the same.

8.4 CRACK WIDTH AND CORROSION LAYER THICKNESS

8.4.1 Relationship Between Crack Width, W_i, and Corrosion Layer Thickness, T_{CL}

Substituting the regression equations of a_1 (Table 8.3) and a_2 (Table 8.4) into Eq. (8.5), the total circumferential crack widths at any radius with

TABLE 8.4 Linear Regression Results for a_2 Compared to T_{CL}

Specimens	Linear Fitting Formula	R^2
R000	$a_2 = 2.923T_{CL}$	0.436
R033	$a_2 = 2.951T_{CL}$	0.658
R067	$a_2 = 2.973T_{CL}$	0.768
R100	$a_2 = 2.987T_{CL}$	0.736

TABLE 8.5 Substituting Results of W_i and Critical Crack Width W_c

Specimens	W_i (mm)	W_c (mm)
R000	$W_i = (0.0919T_{CL} - 0.0029)l_{ci} + 2.923T_{CL}$	0.092
R033	$W_i = (0.0931T_{CL} - 0.0028)l_{ci} + 2.951T_{CL}$	0.089
R067	$W_i = (0.0981T_{CL} - 0.0021)l_{ci} + 2.973T_{CL}$	0.064
R100	$W_i = (0.1025T_{CL} - 0.0014)l_{ci} + 2.987T_{CL}$	0.041

different corrosion layer thicknesses for R000, R033, R067, and R100 are expressed in Table 8.5. W_c represents the critical crack width at the moment of concrete surface cracking that is discussed in Section 8.4.2.

Based on the aforementioned equations, the relationship between W_i, which is the crack width at any distance from the radius R_i to the surface of the steel bar, and the corrosion layer thickness, T_{CL}, can be expressed. Using R000 as an example, the total crack width on the concrete surface (when $l_{ci} = 30$ mm) can be expressed as:

$$W_s = W_{i=30} = 5.680T_{CL} - 0.087 \tag{8.6}$$

whereas the total crack width on the steel surface (when $l_{ci} = 0$) is:

$$W_{steel} = W_{i=0} = 2.923T_{CL} \tag{8.7}$$

Substituting the critical corrosion layer thickness at the surface cracks of the concrete cover (ie, 0.0316 mm for R000) into Eqs. (8.6) and (8.7), 0.092 mm is calculated from both equations. This result indicates that the crack width at the moment of concrete surface cracking, W_c, remains constant along the radial direction at the moment of concrete surface cracking.

Similar results were obtained for specimens R033, R067, and R100 (ie, the crack widths calculated from Eqs. (8.6) and (8.7) were mostly the same), with values of 0.089, 0.064, and 0.041 mm, respectively.

8.4.2 W_c: Critical Crack Width at Concrete Outer Surface Cracking

In previous studies, the crack width was normally defined as 0 when calculating the critical steel corrosion at surface cracking of the concrete cover. However, based on the statement in Sections 8.3.2 and 8.4.1, the crack width cannot be 0 at the time of concrete surface cracking; it maintains the same value along the radial direction in the concrete cover. Therefore, it is inaccurate to predict the steel corrosion at concrete surfacing cracking by allowing the surface crack width to equal 0. A nonzero crack width should be introduced when predicting the steel corrosion at surface cracking.

In this work, the values of the critical crack widths were 0.092, 0.089, 0.064, and 0.041 mm, respectively, for the R000, R033, R067, and R100 specimens listed in Table 8.5. It can be observed that the critical crack width at the concrete surface cracking decreases with the increase of the RA replacement percentage in the concrete mixture. Because there are more weak interfaces in RAC [1–4], the cracks would generate at these interfaces with the development of the corrosion layer, thus leading to the timely release of the tensile stress in the concrete cover. Moreover, the corrosion-induced crack width is proportional to the steel corrosion amount [5–15] because the critical corrosion layer thickness decreases with the increase of RA replacement, as shown in Table 8.3, and the crack width of the RAC at concrete surface cracking, which corresponds to the critical corrosion layer thickness, decreases with the increase of RA replacement as well.

8.4.3 W_s: Crack Width on the Surface of Concrete Cover

The measured surface crack widths, W_s, are listed in Table 8.2. Fig. 8.9 plots the measured surface crack widths, W_s, compared to the corrosion layer thickness, T_{CL}. It can be observed from the figure that the surface crack width, W_s, increases linearly with the growth of the corrosion layer thickness, T_{CL}. The linear regression equations for the four types of concrete specimens are provided in Table 8.6. It should be noted that in these regression equations, the values of the critical corrosion layer thickness, $T_{CL}^{surface}$ (Table 8.3), corresponds to the crack widths, W_c (Table 8.5), representing the situation of the concrete surface cracking moment.

It also can be found that the slope of the regression line increases with the increased RA replacement percentage. This result indicates that, at the same corrosion layer thickness, increasing the RA content in the concrete leads to a larger crack width on the surface of the concrete cover. This is primarily caused by the inferior quality of the RAC [1–4].

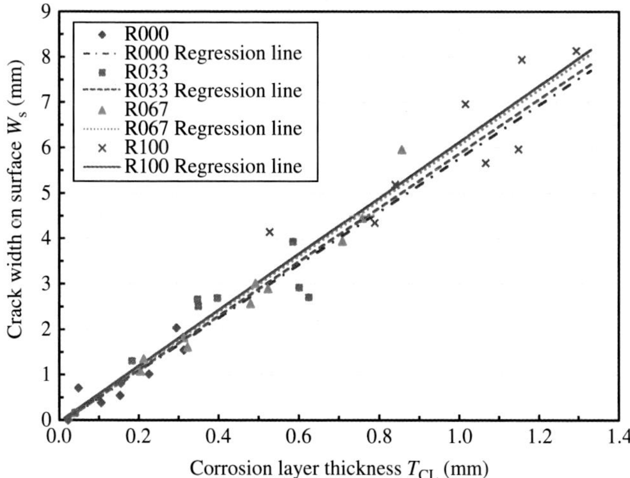

FIGURE 8.9 Relationship between crack width on concrete surface, W_s, and corrosion layer thickness, T_{CL}.

TABLE 8.6 Linear Regression Results of W_s Compared to T_{CL}

Specimens	Regression Results	R^2
R000	$W_s = 5.8528T_{CL} - 0.0929$	0.8385
R033	$W_s = 5.9566T_{CL} - 0.0903$	0.8163
R067	$W_s = 6.1105T_{CL} - 0.0668$	0.9530
R100	$W_s = 6.1597T_{CL} - 0.0434$	0.7337

8.5 RELATIONSHIP OF CORROSION LAYER THICKNESS T_{CL} AND CRACK WIDTH VARIATION COEFFICIENT a_1

To discuss the general relationship of T_{CL} and a_1, R000 is used as an example; Fig. 8.10 illustrates the relationship of T_{CL} and a_1 in specimen R000. The regression line of the R000 data intersects with the x-axis at the critical corrosion layer thickness of $T_{CL} = 0.0316$ mm, where $a_1 = 0$.

If $T_{CL} < 0.0316$ mm, then a_1 is negative, meaning that the crack width near the surface of the concrete cover is smaller than that near the steel bar. This occurs prior to concrete surface cracking. At this moment, the crack width linearly decreases to zero from the surface of the steel bar to the surface of the concrete cover according to the model, as shown in Fig. 8.10 (before surface cracking). In this study, slice R000-1-8 ($T_{CL} = 0.0229$ mm),

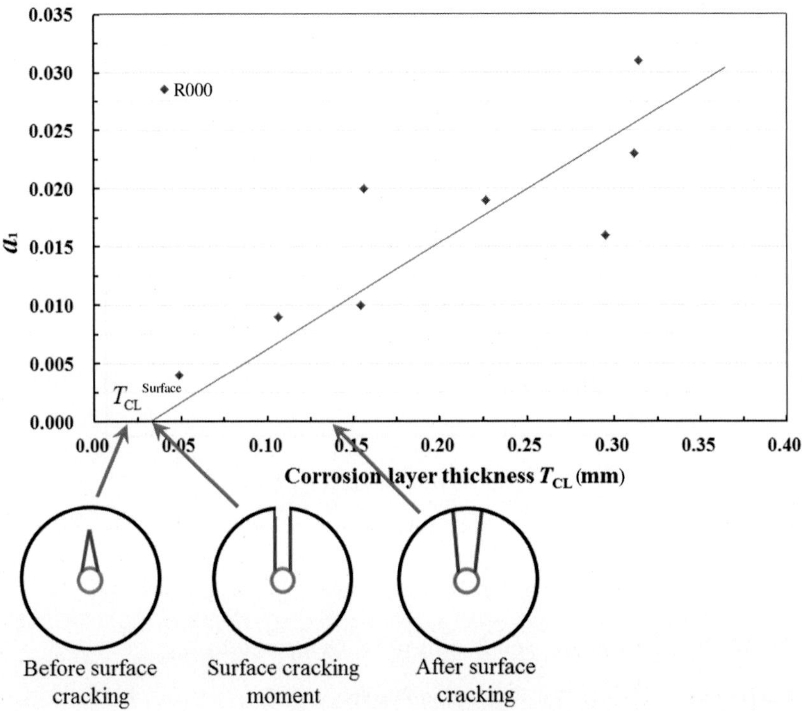

FIGURE 8.10 Schematic diagram of corrosion-induced crack propagation.

as shown in Fig. 8.5a, reflects this situation, the fitting result of which is $a_1 = -0.002$.

When the corrosion layer thickness, T_{CL}, reaches the critical value 0.0316 mm, a_1 equals 0, physically means that the crack width is constant along the radial direction, as shown in Fig. 8.10 (surface cracking moment). This is assumed as the moment when concrete surface cracking occurs; the corrosion layer thickness at this moment is defined as $T_{CL}^{surface}$, as discussed in Section 8.3.2.

If $T_{CL} > 0.0316$ mm, then a_1 is positive, meaning that the crack width near the surface of the concrete cover is larger than that near the steel bar, as shown in Fig. 8.10 (after surface cracking). This occurs after the crack has penetrated the concrete cover. In this study, most of the slices exhibited this situation, as shown in Fig. 8.5b.

Based on this, it appears that, with the increase of steel corrosion, the inner crack propagates outward until it reaches the outer surface of the concrete cover. Correspondingly, parameter a_1 changes from a negative to zero at concrete surface cracking and then becomes positive afterward.

The phenomenon was also observed in the other three types of RAC specimens.

8.6 CRACK SHAPE IN DIFFERENT TYPES OF CONCRETE

Although the corrosion-induced crack propagation processes of different types of concrete (NAC and RAC in this study) are similar, as presented in Section 8.3.1, the corrosion-induced crack shapes of NAC and RAC do have differences. Based on the parameters in the model, which was proposed to describe the variations of the crack width along the radial direction in Section 8.3, the differences in crack shapes of RAC and NAC when corrosion-induced cracks propagate is summarized in Table 8.7.

Prior to surface cracking, only inner cracks exist. At the surface of the steel bar (where $l_{ci} = 0$), $W_{i=0} = a_2$. Because the value of a_2 for the four types of concrete are very close (Section 8.3.3), the crack shapes of the inner cracks for both NAC and RAC are mostly the same.

At the surface cracking moment, when $a_1 = 0$, $T_{CL} = T_{CL}^{surface}$, and $W_i = W_c$, the crack width maintains the same value along the radial direction in the concrete cover (Section 8.4.1). The critical corrosion layer thickness $T_{CL}^{surface}$ of the RAC is smaller than that of the NAC, as presented in Section 8.4.2, and $T_{CL}^{surface,NAC} > T_{CL}^{surface,RAC}$. Therefore, it is easier for RAC to crack than it is for NAC to crack. In addition, the critical width, W_c, at the concrete outer surface cracking decreases with the replacement of RA, as discussed in Section 8.4.2. Therefore, it can be concluded that $W_c^{NAC} > W_c^{RAC}$ at the moment of concrete surface cracking induced by steel corrosion.

After surface cracking, the crack width increases with the radius and corrosion layer thickness. For a given corrosion layer thickness, the angle of cracks γ in the RAC is larger than that of the NAC, as reflected by a_1 (Section 8.3.2). Therefore, the propagation of crack width in the RAC was more rapid than that of the NAC, and the crack widths on the surface of the concrete cover of the RAC are larger than those of the NAC when corresponding to the same corrosion layer thickness value: $\gamma^{NAC} < \gamma^{RAC}$ and $W_s^{NAC} < W_s^{NAC}$.

Based on these investigations, the increased RA in the concrete mixture not only results in the early corrosion-induced concrete surface cracking but also accelerates the width development of the crack. This is because the anti-cracking properties of the RAC are weaker than those of the NAC. Because the crack shape differences between RAC and NAC were primarily caused by the concrete quality, differences can also apply to concrete with different qualities.

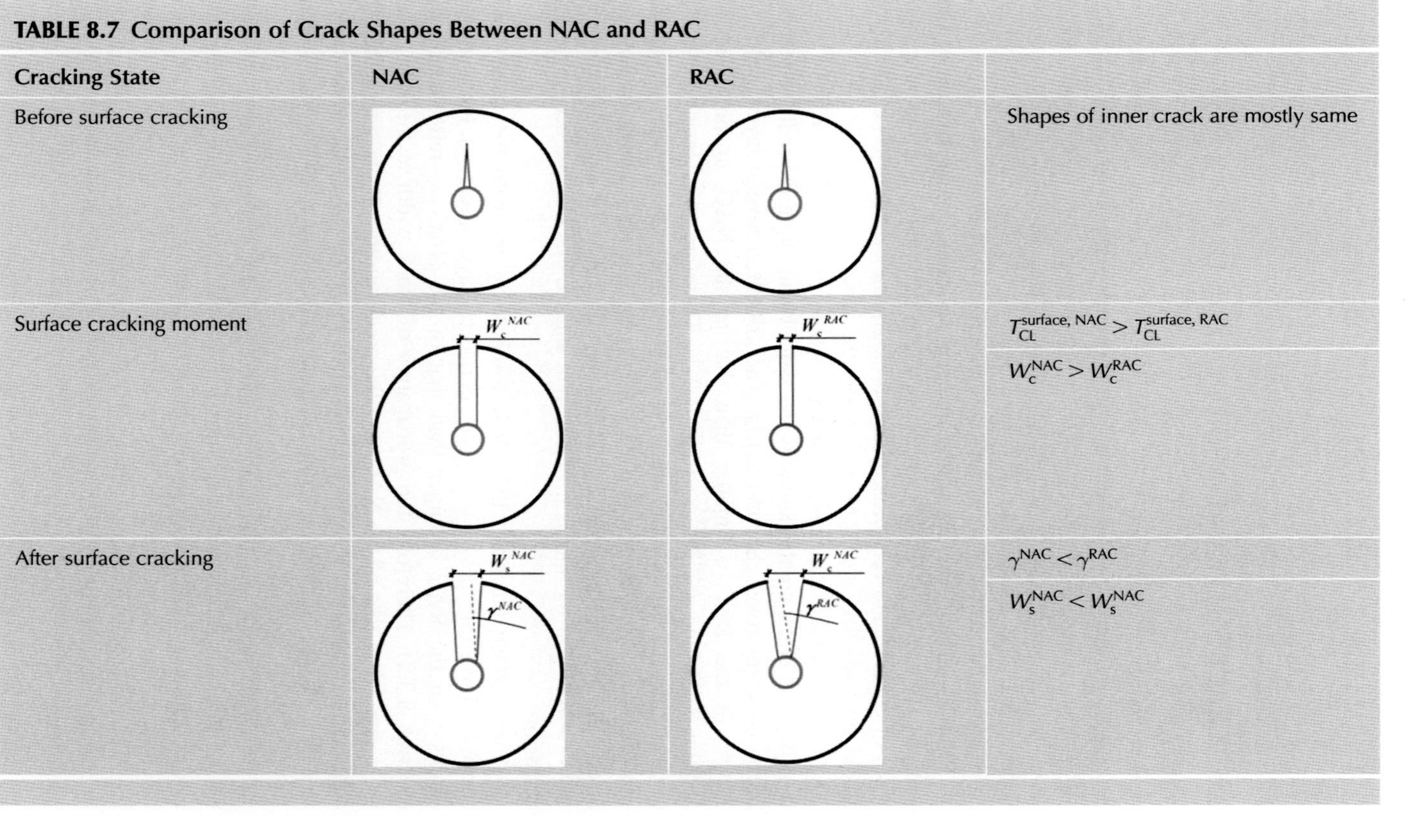

TABLE 8.7 Comparison of Crack Shapes Between NAC and RAC

Cracking State	NAC	RAC	
Before surface cracking			Shapes of inner crack are mostly same
Surface cracking moment			$T_{CL}^{\text{surface, NAC}} > T_{CL}^{\text{surface, RAC}}$
			$W_c^{NAC} > W_c^{RAC}$
After surface cracking			$\gamma^{NAC} < \gamma^{RAC}$
			$W_s^{NAC} < W_s^{NAC}$

8.7 CONCLUSIONS

1. Linear models were proposed to describe the variation of the total circumferential crack width along the radial direction. Parameters a_1 and a_2 in the model can be interpreted physically to describe various characteristics of the cracks. The physical meaning of a_1 is the crack width variation coefficient, and that of a_2 is the crack width coefficient at the surface of the steel bar.
2. The corrosion-induced crack propagation process was discussed and the crack shape prior to, at the point of, and after the concrete surface cracking were schematically described. The differences between the corrosion-induced crack shapes of different types of concrete were also discussed.
3. At the concrete surface cracking moment, the crack width in the concrete cover is not zero, it maintained a constant width along the radial direction.
4. Worse concrete (RAC in this study) quality not only resulted in early corrosion-induced concrete surface cracking but also accelerated the propagation of the cracks in concrete cover.

REFERENCES

[1] Gómez-Soberón J. Porosity of recycled concrete with substitution of recycled concrete aggregate: an experimental study. Cement Concrete Res 2002;32(8):1301–11.

[2] Tam VW, Gao XF, Tam CM. Microstructural analysis of recycled aggregate concrete produced from two-stage mixing approach. Cement Concrete Res 2005;35(6):1195–203.

[3] Poon CS, Shui ZH, Lam L. Effect of microstructure of ITZ on compressive strength of concrete prepared with recycled aggregates. Constr Build Mater 2004;18(6):461–8.

[4] Xiao JZ. Recycled concrete. Beijing, P.R. China: Chinese Building Construction Publishing; 2008 [in Chinese]

[5] Alonso C, Andrade C, Rodriguez J, et al. Factors controlling cracking of concrete affected by reinforcement corrosion. Mater Struct 1998;31(7):435–41.

[6] Oh BH, Kim KH, Jang BS. Critical corrosion amount to cause cracking of reinforced concrete structures. ACI Mater J 2009;106(4):333–9.

[7] Vidal T, Castel A, Francois R. Analyzing crack width to predict corrosion in reinforced concrete. Cement Concrete Res 2004;34(1):165–74.

[8] Vu K, Stewart MG, Mullard J. Corrosion-induced cracking: experimental data and predictive models. ACI Struct J 2005;102(5):719–26.

[9] Mullard JA, Stewart MG. Corrosion-induced cover cracking of RC structures: new experimental data and predictive models NOVA. The University of Newcastle's Digital Repository[J]; 2009.

[10] Zhang R, Castel A, François R. Concrete cover cracking with reinforcement corrosion of RC beam during chloride-induced corrosion process. Cement Concrete Res 2010;40 (3):415–25.

[11] Liu Y, Weyers RE. Modeling the time-to-corrosion cracking in chloride contaminated reinforced concrete structures. ACI Mater J 1998;95(6):675–81.

[12] Andrade C, Alonso C, Molina FJ. Cover cracking as a function of bar corrosion: part I—experimental test. Mater Struct 1993;26(8):453–64.

[13] Rasheeduzzafar, Al-Saadoun SS, Al-Gahtani AS. Corrosion cracking in relation to bar diameter, cover, and concrete quality. J Mater Civil Eng 1992;4(4):327–42.

[14] Bhargava K, Ghosh AK, Mori Y, et al. Modeling of time to corrosion-induced cover cracking in reinforced concrete structures. Cement Concrete Res 2005;35(11):2203–18.

[15] Zhao YX, Jin WL. Modeling the amount of steel corrosion at the cracking of concrete cover. Adv Struct Eng 2006;9(5):687–96.

Chapter 9

Development of Corrosion Products-Filled Paste at the Steel–Concrete Interface

Chapter Outline

9.1 INTRODUCTION

In chapter "Crack Shape of Corrosion-Induced Cracking in the Concrete Cover," steel corrosion accumulating at the steel–concrete interface is called the corrosion layer (CL) and the concrete zone around the steel bar filled by corrosion products is called corrosion products-filled paste (CP). The CL and CP are used to describe these two regions in this chapter.

With respect to the development of CP and CL, the previous research assumed that the development of CL falls behind the development of CP [1–4]. Quite a few studies [5–12] have proven the existence of CP, but more quantitative experimental research is needed to study the thickness of CP and its variation with the growth of steel corrosion. The authors have studied CP in chapter "Rust Distribution in Corrosion-Induced Cracking Concrete" and found that the average CP thickness increases as the CL thickness grows. This result shows that the penetration of corrosion products into the porous zone of concrete and the formation of a CL at the steel–concrete interface proceed simultaneously, not separately (as previously assumed), after the initiation of steel corrosion.

Because the thickness of CP and CL and their relationship are essential for the prediction of concrete surface cracking induced by steel corrosion, and considering the limitation of the specimens used in previous work stated

Steel Corrosion-Induced Concrete Cracking. DOI: http://dx.doi.org/10.1016/B978-0-12-809197-5.00009-8

in chapter "Rust Distribution in Corrosion-Induced Cracking Concrete," this chapter investigates the development of CL and CP in four types of corroding concrete specimens and provides quantitative descriptions of the relationship between T_{CP} and T_{CL}.

The specimens and their exposure history are the same as used in chapter "Crack Shape of Corrosion-Induced Cracking in the Concrete Cover." The sample preparation and the measurements of CP and CL are the same as stated in Section 5.2.4. For more information regarding the specimens and the measurements, please refer to chapter "Crack Shape of Corrosion-Induced Cracking in the Concrete Cover." Approximately 10 slices of each specimen were produced in total.

9.2 INFLUENCE OF CRACKS ON CP THICKNESS

Fig. 9.1 is a BSE image at the steel–concrete interface. From Fig. 9.1a–c, a sudden increase of T_{CP} near the inner crack can be seen, which is defined as a crack not penetrating the concrete cover. This is a typical phenomenon reflecting the influence of inner cracks on CP development. The authors

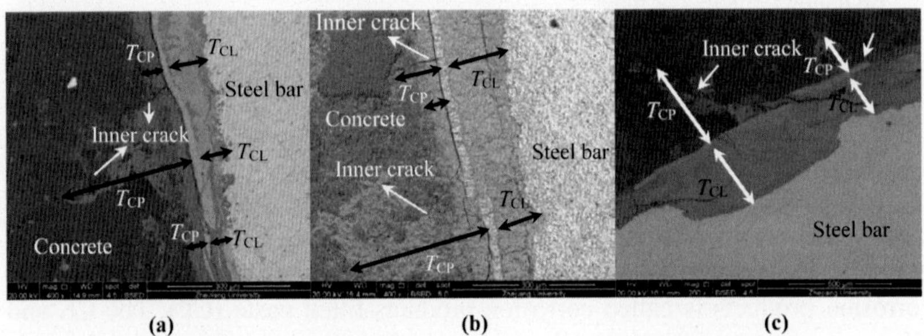

Once the inner crack develops outwards and penetrates the concrete cover, the solution containing Ferric ions transports outside directly, there is then no extra increase of CP thickness in the outer cracks.

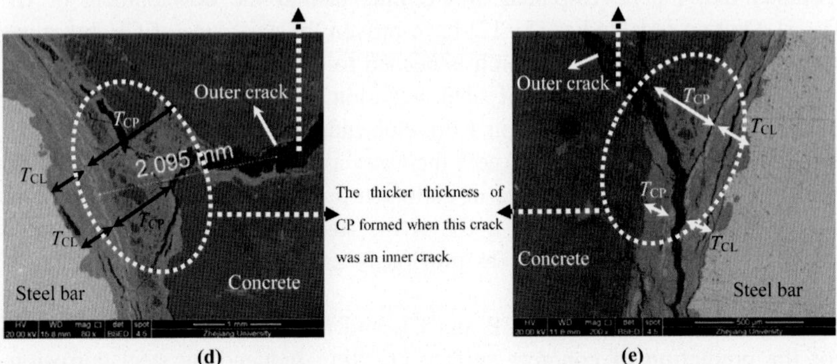

FIGURE 9.1 Influence of cracks on CP development. (a–c) Inner crack. (d, e) Outer crack. *CP*, corrosion products-filled paste; *CL*, corrosion layer; T_{CP}, thickness of CP; T_{CL}, thickness of CL.

assume that this phenomenon is mainly induced by the accelerated corrosion history of the wetting and drying cycles. During the wetting period, the electrons flow between the stainless steel nets and the steel bar, and the moisture in the concrete cover provides favorable conditions for the ferric ions to migrate outward from the steel surface, particularly through the easy paths (ie, the inner cracks). During the drying period, the solution containing ferric ions in the inner cracks penetrates the adjacent concrete surfaces, resulting in a sudden increase in the thickness of the CP layer. It needs be pointed out that the inner crack causes "a sudden increase" of T_{CP} in most images; however, as shown in Fig. 9.1a, there is no obvious increase in T_{CP} near the "upper short inner crack." This phenomenon is regarded as a special case. The upper short inner crack in Fig. 9.1a might occur just before the corrosion acceleration process is stopped, and there is not enough time for the ferric ions to migrate outward through this upper short inner crack.

With the steel corrosion increasing, the inner cracks penetrate the depth of the concrete cover propagating outward to the concrete surface (ie, developing into outer cracks), which can provide a path for the solution containing ferric ions to be directly transported outside instead of residing in the concrete cracks. Therefore, there is no extra increase of CP thickness in the outer cracks. As shown in Figs. 9.1d and e, it is typical for the thickness of CP around the inside part of the outer crack to be larger; however, far away from the steel bar, there is no sign of CP existence. It can be inferred that CP around the inside part of the outer crack formed when the crack was still an inner crack.

It needs to be noted that this phenomenon might occur in specimens in environments where wetting and drying occur, such as tidal zones or splash zones; if the specimens were corroded in the atmospheric environment, without the solution ingression, then this phenomenon might not occur.

9.3 RELATION BETWEEN T_{CP} AND T_{CL} EXCLUDING THE EFFECT OF INNER CRACKS

As stated in Section 9.2, the inner crack causes a sudden increase of T_{CP}. To exclude this effect of the inner crack, the thicknesses of CP and CL at the regions shown in Fig. 9.2 are measured, avoiding the areas at the inner cracks.

Fig. 9.3 illustrates the relationship between the measured data of T_{CP} and T_{CL} from all of the R000 samples. The measured data in Fig. 9.3a are scattered and cannot provide a clear relationship between T_{CP} and T_{CL}. To find the intrinsic relationship between T_{CP} and T_{CL}, the statistical results of the data from all samples were observed because they display some regularity when these scattered data are graded into several groups in the range of 100 μm, as shown in Fig. 9.3b.

It needs to be pointed out that the authors actually tried to group the measured data in the ranges of 20, 40, 50, 100, and 200 μm. However, the

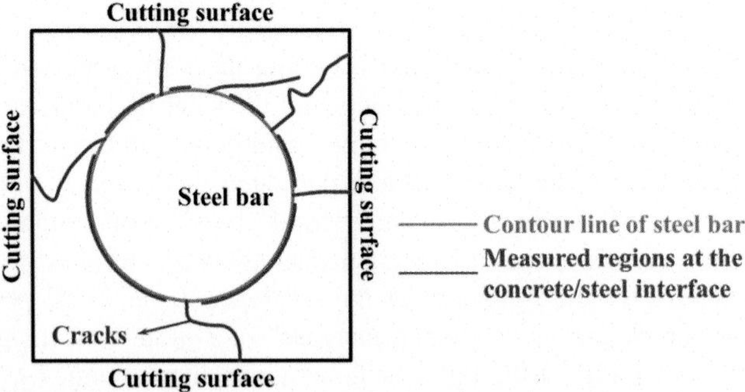

FIGURE 9.2 Schematic of measured regions at the concrete–steel interface.

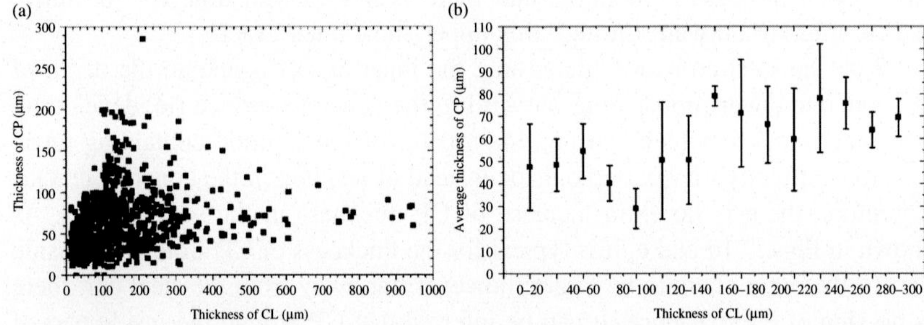

FIGURE 9.3 Thickness of the corrosion products-filled paste (CP) versus thicknesses of the corrosion layer (CL) excluding the regions of the inner cracks for R000. (a) Measured data. (b) A part of the data map after grouping in the range of 20 μm for R067.

results of grouping in the ranges of 20, 40, 50, and 200 μm are not able to provide the information as clearly as results of grouping in the range of 100 μm, which display good regularity between T_{CP} and T_{CL}. Therefore, in this study, the data are grouped in the range of 100 μm; this is the same grouping method as that used by the authors in chapter "Rust Distribution in Corrosion-Induced Cracking Concrete."

As illustrated in Fig. 9.3b, the trend is that as the T_{CL} increases, the average T_{CP} develops gradually until the T_{CL} is up to a certain value. After that, the average T_{CP} remains nearly unchanged.

The same data analysis process was conducted for R033, R067, and R100. Fig. 9.4 illustrates the results for four types of the specimens. It can be seen that the tendency for all specimens is similar: the average CP thickness increases as the CL thickness grows; when CL thickness reaches a certain value, the average CP thickness has no obvious growth trend. This result proves that the proposed T_{CP}–T_{CL} model, as presented in chapter "Rust Distribution in Corrosion-Induced Cracking Concrete," is reasonable.

It also can be seen from Fig. 9.4 that T_{CL}^{cr} is approximately 180 μm, regardless of the concrete type. That is, for all specimens, as the T_{CL}

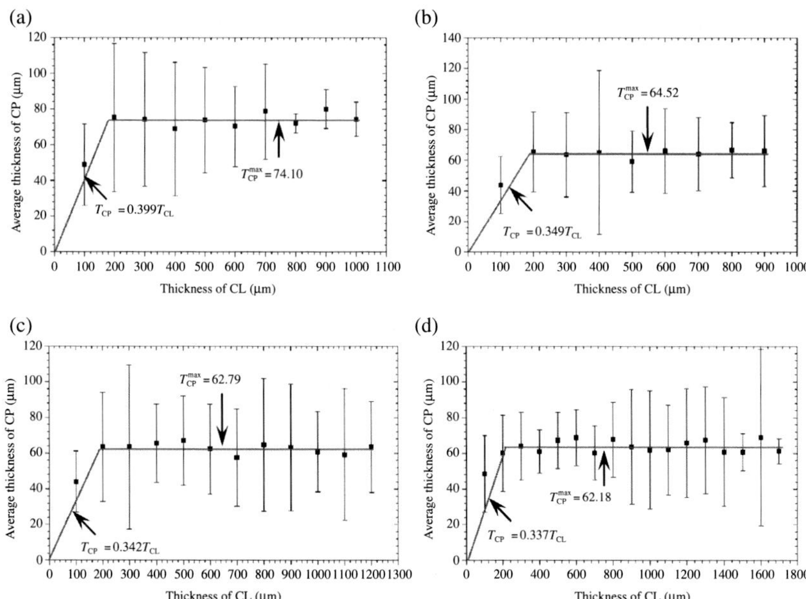

FIGURE 9.4 Relationship between T_{CP} and T_{CL} excluding the regions of the inner cracks. (a) R000. (b) R033. (c) R067. (d) R100.

TABLE 9.1 Fitting Values of k_T and T_{CP}^{max} (μm)

Concrete Type	R000	R033	R067	R100
k_T	0.399	0.349	0.342	0.337
T_{CP}^{max}	74.10	64.52	62.79	62.18

increases, the average T_{CP} develops gradually until the T_{CL} reaches 180 μm, and then T_{CP} remains unchanged. The fitting values of k_T and T_{CP}^{max} in Fig. 9.4 are presented in Table 9.1, where k_T is the ratio between T_{CP}^{max} and T_{CL}^{cr}, T_{CP}^{max} is the maximum of the thickness of CP.

To compare the T_{CP}–T_{CL} relationship of four types of concrete specimens, the fitting lines are presented in one figure, as illustrated in Fig. 9.5. It can be easily seen from Fig. 9.5 that, with the increase of the RA replacement, the values of k_T and T_{CP}^{max} of four types of concrete specimens decrease.

When the replacement percentage of RA increases, the quality of concrete decreases because the porosity of RAC increases [13–16]. This quality degradation can also be proven by the decrease of the compressive strength with the increase of the RA replacement listed in Table 8.1. It can be seen

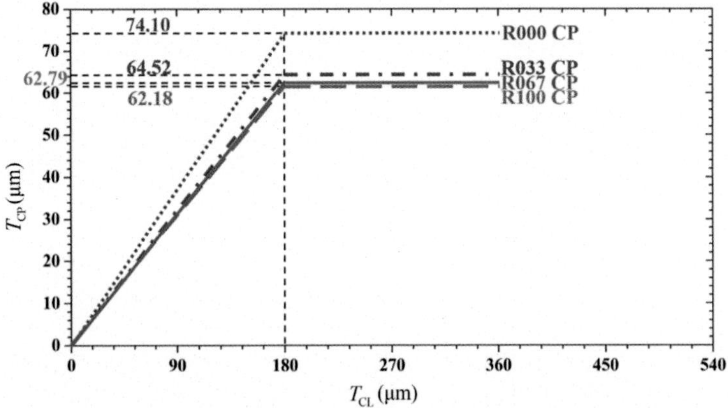

FIGURE 9.5 T_{CP}–T_{CL} models for four types of concrete (excluding the effect of inner cracks).

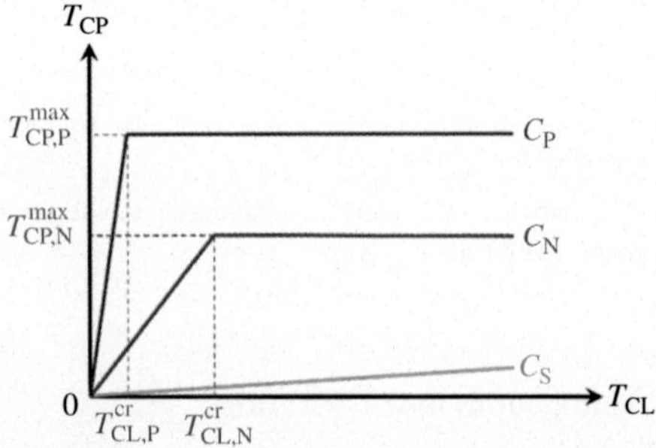

FIGURE 9.6 Effect of concrete quality on T_{CP}.

from Fig. 9.5 that the worst quality concrete R100 has the lowest values of k_T and T_{CP}^{max}, whereas the highest quality concrete R000 gives the highest values of k_T and T_{CP}^{max}.

The T_{CP}–T_{CL} model is discussed here with regard to three different concrete types, including one normal concrete and two extreme cases, as shown in Fig. 9.6. The normal concrete is denoted as C_N. One extreme type is very solid concrete with extremely low porosity, denoted as C_S, whereas the other is a very porous concrete, denoted as C_P. For the concrete, C_S, with high quality, it is difficult for rust to fill the concrete; therefore, as T_{CL} increases, T_{CP} remains nearly zero, as illustrated in Fig. 9.6. For the concrete, C_P, a large part of the corrosion product generated by the steel fills the pores because there are quite a few pores in the concrete adjacent to the steel bar; therefore, the T_{CL} increases slowly while the T_{CP} grows significantly, resulting in the large value of k_T for C_P. For the normal concrete, C_N, the porosity of C_N is between these two extreme cases, and the values of k_T and T_{CP}^{max} of C_N are also in between, as shown in Fig. 9.6.

Comparing Fig. 9.5 and Fig. 9.6, it is clear that the tendency of the four types of concrete specimens illustrated in Fig. 9.5 is not in accord with the aforementioned discussion, as shown in Fig. 9.6. This is because the tested regions in this section exclude the inner cracks. The more RA replacement in concrete, the worse the quality of the concrete is, and the easier it is for the inner cracks to occur. Therefore, the replacement concrete with higher RA has more inner cracks, which were also observed during the CP thickness measurement. As stated in Section 9.2, these inner cracks provide a more convenient path for the migration of the ferric ions and the accommodation of steel corrosion products; therefore, the concrete with higher RA replacement has more inner cracks that accommodate more corrosion products, leaving less steel corrosion products in the noncracking areas from which the measured data were obtained. This is the explanation why the values of k_T and T_{CP}^{max} decrease when the RA replacement increases.

However, to predict the corrosion-induced concrete cracking, the corrosion products filling the inner cracks before concrete surface cracking should be considered in the concrete cracking model, and the data analysis excluding the effect of inner cracks might not be in accord with the actual situation. Therefore, the CP measurements considering the inner crack areas are discussed in Section 9.4.

9.4 RELATION BETWEEN T_{CP} AND T_{CL} INCLUDING THE INNER CRACKS

Again take R000 for example; Fig. 9.7a illustrates the relationship between the measured data of T_{CP} and T_{CL} from all of the samples of R000, including the effect of inner cracks. Fig. 9.7b reflects local magnification of Fig. 9.7a. Compared with Fig. 9.3a, it can be seen that the data of T_{CP} and T_{CL} considering the effect of inner cracks in Fig. 9.7b are much messier. The grouping method used in Section 9.3 does not work for the data in this section. Therefore, an alternative method is used for the data analysis here.

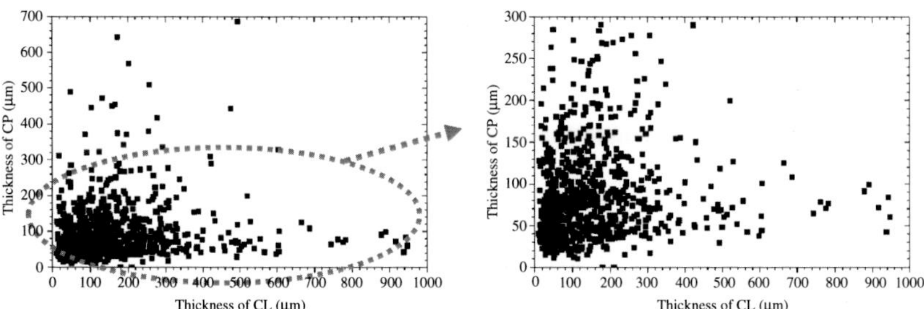

FIGURE 9.7 Thickness of corrosion products-filled paste (CP) versus thickness of corrosion layer (CL) including the regions of the inner crack for R000s. (a) All measured data. (b) Local magnification.

Approximately 100 points, evenly distributed around the rebar, were observed to cover the entire steel–concrete interface, including the inner cracks areas, and the thicknesses of the CL and CP were measured at those points. The average T_{CP} and T_{CL} for each sample are listed in Table 9.2. According to the result in Section 9.3, T_{CL}^{cr} is approximately 180 μm, regardless of the concrete type, as shown in Fig. 9.5. The authors assume that this rule is suitable for the data analysis, including the effect of inner cracks. Therefore, the authors average the measured T_{CP} if their corresponding T_{CL} is near or greater than 180 μm. The mean values of T_{CP} and their mean square deviations are also listed in Table 9.2.

Fig. 9.8 illustrates the relationship between T_{CP} and T_{CL} of each sample; their average lines are also plotted in this figure. The tested T_{CP} data are marked as solid if they were not considered in the averaging calculation (the black bold data in Table 9.2). It can be seen that the average T_{CP} has a gradient increase with the increase of RA replacement, which is opposite to the tendency stated in Section 9.3 but in line with the trend presented in Fig. 9.6. Therefore, the CP data analysis including the inner cracking areas can reflect the actual situation better, although the data of T_{CP} near the inner cracks are messy.

Based on the aforementioned analysis, the $T_{CP} - T_{CL}$ models for four types of concrete specimens considering the effect of inner cracks are proposed in Fig. 9.9. It needs to be noted that using 180 μm as T_{CL}^{cr} is based on the observation of this study; whether 180 μm is widely suitable for other types of concrete as T_{CL}^{cr} still needs more experimental investigation.

According to Fig. 9.9, the values of k_T considering the effect of inner cracks of R000, R033, R067, and R100 can be easily calculated as listed in Table 9.3.

The values of k_T considering the effect of inner cracks of R000, R033, R067, and R100 increase with the increase of RA replacement. This is because when the replacement percentage of RA increases, the quality of concrete decreases, resulting in the larger k_T. In chapter "Rust Distribution in Corrosion-Induced Cracking Concrete," the value of k_T of the high-performance concrete is approximately 0.3, which is smaller than the k_T of the normal and RA concrete in this study. This is in line with the trend shown in Fig. 9.6 because the quality of the high-performance specimens in the previous work is much better than that of the four types of concrete specimens used in this work.

In comparison with the analysis in Section 9.3, the $T_{CP} - T_{CL}$ model established for the data considering the effect of the inner cracks is more realistic and suitable for the prediction of corrosion-induced concrete surface cracking. However, as stated, more experimental work needs to be performed considering the different concrete mix proportions and the various service conditions to improve the $T_{CP} - T_{CL}$ model.

TABLE 9.2 Average T_{CP} and T_{CL} of Each Sample, Mean Values of T_{CP}, and Their Mean Square Deviation of Each Type of Concrete Specimens (μm)

Sample	T_r	T_{CL}	T_{CP}	Mean Values of T_{CP}*	Mean Square Deviation
R000-1-5	314.1	247.11	97.23		
R000-2-3	153.9	175.72	98.32		
R000-2-5	295.4	209.52	104.83	99.28	2.85
R000-2-6	226.7	192.33	98.81		
R000-2-7	312.0	217.45	97.19		
R000-1-4	**48.9**	**78.19**	**83.12**		
R000-1-6	**106.1**	**106.96**	**89.27**		
R000-1-7	**68.6**	**91.94**	**86.90**		
R000-2-8	**156.0**	**149.02**	**94.46**		
R033-1-4	602.2	410.77	103.41		
R033-1-5	348.4	178.06	102.60	103.49	0.76
R033-1-6	184.1	160.23	104.46		
R033-1-7	**40.6**	**34.88**	**60.68**		
R033-1-8	**41.6**	**34.59**	**60.79**		
R067-1-1	323.4	281.28	119.75		
R067-2-3	857.0	833.03	145.47		
R067-2-4	709.3	720.22	150.91		
R067-2-5	524.2	532.36	126.78	130.37	14.94
R067-2-6	493.0	467.95	133.98		
R067-2-7	480.5	316.73	133.00		
R067-2-8	314.1	173.29	102.68		
R067-1-2	**204.9**	**119.98**	**109.06**		
R067-1-3	**213.2**	**118.95**	**98.62**		
R100-1-3	1016.1	664.70	172.61		
R100-1-4	1157.5	1040.88	192.34		
R100-1-5	1294.8	1337.35	200.90		
R100-1-6	1149.2	969.08	171.72	180.28	15.00
R100-1-7	1067.0	729.89	181.16		
R100-1-8	789.4	671.80	190.60		
R100-1-9	528.3	357.82	152.63		

* The mean value of T_{CP} is the average of T_{CP} whose corresponding T_{CL} is near or greater than 180 μm.

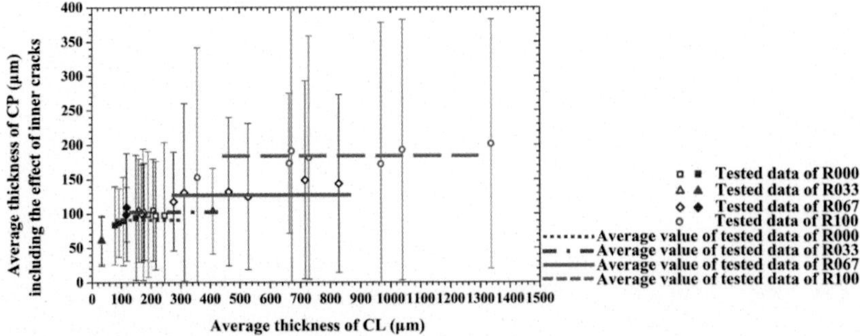

FIGURE 9.8 Tested data of all samples and their average value of tested data.

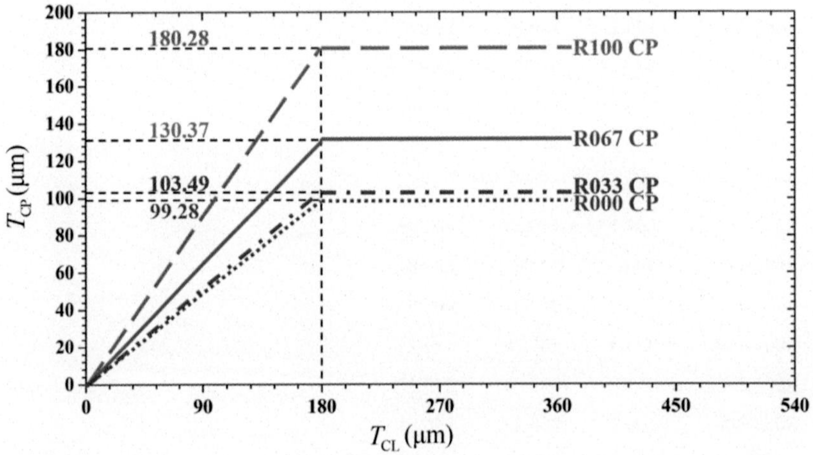

FIGURE 9.9 T_{CP}–T_{CL} models for four types of concrete (including the effect of inner cracks). CP, corrosion products-filled paste; T_{CP}, thickness of CP; T_{CL}, thickness of corrosion layer.

TABLE 9.3 Values of k_T Considering the Effect of Inner Cracks

Concrete Type	R000	R033	R067	R100
k_T	0.525	0.575	0.708	1.002

It needs to be pointed out that the critical thickness of CL at concrete surface cracking reported in the previous research work is normally less than 100 μm, which is smaller than the critical CL thickness when CP is constant, approximately 180 μm in this work. Therefore, it can be assumed that the CP thickness did not reach the maximum value before the concrete surface cracked (ie, somewhere in the sideling, or oblique, line in Fig. 9.9). This means that the penetration of corrosion products into the porous zone of concrete and the formation of a CL at the steel–concrete interface proceed simultaneously before concrete surface cracking; therefore, so-called stage 1 and stage 2 in the corrosion-induced concrete cracking model should not be separately considered, and the cracking model needs to be modified. The improved corrosion-induced concrete surface cracking model is presented in chapter "Corrosion-Induced Concrete Cracking Model Considering Corrosion Products-Filled Paste."

9.5 CONCLUSIONS

1. Inner cracks provide a convenient path for corrosion rust to penetrate to the concrete near the cracks, causing a sudden increase in T_{CP}. However, once the inner crack develops outward and penetrates the concrete cover and the solution containing ferric ions transports directly to the outside, there is no extra increase in CP thickness in the outer cracks.
2. As the T_{CL} grows, the T_{CP} develops gradually until the T_{CL} reaches a certain value (T_{CL}^{cr}), which is approximately 180 μm, regardless of concrete type in this study. After that, the average T_{CP} remains nearly unchanged. This tendency proves that the previously proposed $T_{CP} - T_{CL}$ model in chapter "Rust Distribution in Corrosion-Induced Cracking Concrete" is reasonable.
3. For the $T_{CP} - T_{CL}$ model established for the data including the effects of inner cracks, the values of k_T and T_{CP}^{max} of concrete specimens increase with the growth of the RA replacement.

REFERENCES

[1] Liu Y, Weyers RE. Modeling the time-to-corrosion cracking in chloride contaminated reinforced concrete structures. ACI Mater J 1998;95(6):675–81.

[2] Chitty W, Dillmann P, L'Hostis V, et al. Long-term corrosion resistance of metallic reinforcements in concrete—a study of corrosion mechanisms based on archaeological artefacts. Corros Sci 2005;47(6):1555–81.

[3] Lu C, Jin W, Liu R. Reinforcement corrosion-induced cover cracking and its time prediction for reinforced concrete structures. Corros Sci 2011;53(4):1337–47.

[4] Kim KH, Jang SY, Jang BS, et al. Modeling mechanical behavior of reinforced concrete due to corrosion of steel bar. ACI Mater J 2010;107(2):106–13.

[5] Zhao YX, Wu YY, Jin W. Distribution of millscale on corroded steel bars and penetration of steel corrosion products in concrete. Corros Sci 2013;66:160–8.

[6] Zhao YX, Yu J, Wu YY, et al. Critical thickness of rust layer at inner and out surface cracking of concrete cover in reinforced concrete structures. Corros Sci 2012;59:316–23.

[7] Wong HS, Zhao YX, Karimi AR, et al. On the penetration of corrosion products from reinforcing steel into concrete due to chloride-induced corrosion. Corros Sci 2010;52(7): 2469–80.

[8] Michel A, Pease BJ, Geiker MR, et al. Monitoring reinforcement corrosion and corrosion-induced cracking using non-destructive X-ray attenuation measurements. Cement Concrete Res 2011;41(11):1085–94.

[9] Duffó GS, Morris W, Raspini I, et al. A study of steel rebars embedded in concrete during 65 years. Corros Sci 2004;46(9):2143–57.

[10] Asami K, Kikuchi M. In-depth distribution of rusts on a plain carbon steel and weathering steels exposed to coastal–industrial atmosphere for 17 years. Corros Sci 2003;45(11): 2671–88.

[11] Caré S, Nguyen QT, L'Hostis V, et al. Mechanical properties of the rust layer induced by impressed current method in reinforced mortar. Cement Concrete Res 2008;38(8): 1079–91.

[12] Zhao YX, Yu J, Hu B, et al. Crack shape and rust distribution in corrosion-induced cracking concrete. Corros Sci 2012;55:385–93.

[13] Olorunsogo FT, Padayachee N. Performance of recycled aggregate concrete monitored by durability indexes. Cement Concrete Res 2002;32(2):179–85.

[14] Etxeberria M, Vázquez E, Marí A, et al. Influence of amount of recycled coarse aggregates and production process on properties of recycled aggregate concrete. Cement Concrete Res 2007;37(5):735–42.

[15] Kou S, Poon C, Etxeberria M. Influence of recycled aggregates on long term mechanical properties and pore size distribution of concrete. Cement Concrete Comp 2011;33(2): 286–91.

[16] Khatib JM. Properties of concrete incorporating fine recycled aggregate. Cement Concrete Res 2005;35(4):763–9.

Chapter 10

Steel Corrosion-Induced Concrete Cracking Model

Chapter Outline

10.1 INTRODUCTION

Through experimental studies of different specimens in chapters "Mill Scale and Corrosion Layer at Concrete Surface Cracking" and "Rust Distribution in Corrosion-Induced Cracking Concrete," the authors have found that corrosion products cannot fill the cracks before the initiation of concrete surface cracking. Therefore, the so-called stage 3 does not need to be considered in the concrete surface cracking model.

According to the experimental results in chapters "Rust Distribution in Corrosion-Induced Cracking Concrete" and "Development of Corrosion Products-Filled Paste at the Steel−Concrete Interface," the average corrosion product-filled paste (CP) thickness increases as the corrosion layer (CL) thickness increases. However, when CL reaches a certain value, the average CP thickness has no obvious growth trend. As has been pointed out in chapter "Development of Corrosion Products-Filled Paste at the Steel−Concrete Interface," the critical thickness of CL at concrete surface cracking reported in the previous research work is normally less than 100 μm, which is smaller than the critical CL thickness when CP is constant; therefore, the CP

Steel Corrosion-Induced Concrete Cracking. DOI: http://dx.doi.org/10.1016/B978-0-12-809197-5.00010-4

thickness did not reach the maximum value before the concrete surface cracked. This means that the penetration of corrosion products into the porous zone of concrete and the formation of a CL at the steel–concrete interface proceed simultaneously before concrete surface cracking. Hence, the so-called stage 1 and stage 2 in corrosion-induced concrete surface cracking model should not be separately considered.

This chapter tries to establish a corrosion-induced concrete surface cracking model that can estimate the corrosion products fill in the paste at the steel–concrete interface and predict the concrete surface cracking more reasonably than the previous models. To achieve this goal, the relationship between the thickness of the CP and the thickness of the CL is modeled based on the experimental work in chapters "Rust Distribution in Corrosion-Induced Cracking Concrete" and "Development of Corrosion Products-Filled Paste at the Steel–Concrete Interface"; the nominal volume expansion rate, n_0, is introduced to consider the effect of corrosion products filling in the paste.

10.2 CORROSION-INDUCED CONCRETE SURFACE CRACKING MODEL CONSIDERING CP

An improved corrosion-induced surface cracking model for concrete is proposed that can calculate the corrosion products filling in concrete pores and accumulating at the steel–concrete interface synchronously.

10.2.1 Cracking Process Description

The cracking process of reinforced concrete caused by steel corrosion is illustrated in Fig. 10.1. After steel depassivation, some steel corrosion products accumulate in the steel–concrete interface to form the CL. The thickness of the CL is T_{CL}. The remainder of the corrosion products fills the voids in the concrete around the steel to form the CP. The thickness of the CP is T_{CP}, as shown in Fig. 10.1b. As the steel corrosion develops, T_{CP}

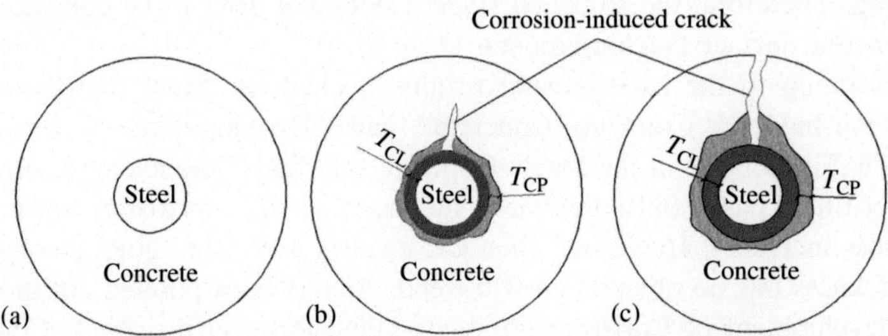

FIGURE 10.1 Corrosion-induced concrete cracking model considering corrosion products-filled paste. (a) Steel depassivation. (b) Corrosion-induced crack appears and CP and CL form simultaneously. (c) T_{CL} and T_{CP} increase gradually until the crack reaches the concrete outer surface.

increases as T_{CL} grows until the crack reaches the outer surface of concrete cover, as in Fig. 10.1c.

The total radial loss of steel when surface concrete cracking can still be calculated with the method presented in chapter "Damage Analysis and Cracking Model of Reinforced Concrete Structures With Rebar Corrosion"; however, the nominal ratio between the corrosion product volume and the basic steel volume, n_0, as stated in the following sections, is introduced for considering the effect of corrosion products filling in the paste.

10.2.2 $T_{CP} - T_{CL}$ Model

Based on the experimental results in chapters "Rust Distribution in Corrosion-Induced Cracking Concrete" and "Development of Corrosion Products-Filled Paste at the Steel–Concrete Interface," the relationship between $T_{CP} - T_{CL}$ and T_{CL} is modeled as shown in Fig. 10.2. Mathematically, this can be expressed as follows:

$$\begin{cases} T_{CP} = k_T \times T_{CL}, & T_{CL} < T_{CL}^{cr} \\ T_{CP} = T_{CP}^{max} = k_T \times T_{CL}^{cr}, & T_{CL} \geq T_{CL}^{cr} \end{cases} \tag{10.1}$$

where T_{CP}^{max} is the maximum T_{CP} achieved, T_{CL}^{cr} is the critical value of CL thickness corresponding to T_{CP}^{max}, and k_T is the ratio between T_{CP}^{max} and T_{CL}^{cr}.

To calculate CP in the concrete cracking model, CP is converted to an evenly distributed CL around the steel. The thickness of the CL of this distributed layer is defined as $T_{CL,pore}$, as shown in Fig. 10.3. It needs to be noted that uniformly distributed steel corrosion is an ideal state; in practical applications, most of the steel bars in concrete corrode nonuniformly. This situation is discussed in Section 10.4.

Assume that in the region of CP the voids are filled completely with corrosion products; then, $T_{CL,pore}$ can be expressed as follows:

$$T_{CL,pore} = \frac{\phi \times \int_0^{2\pi} T_{CP} R d\theta}{2\pi R} \tag{10.2}$$

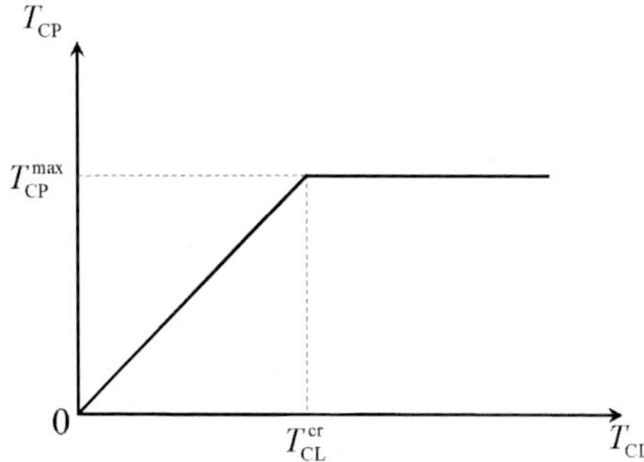

FIGURE 10.2 Relationship between T_{CP} and T_{CL}.

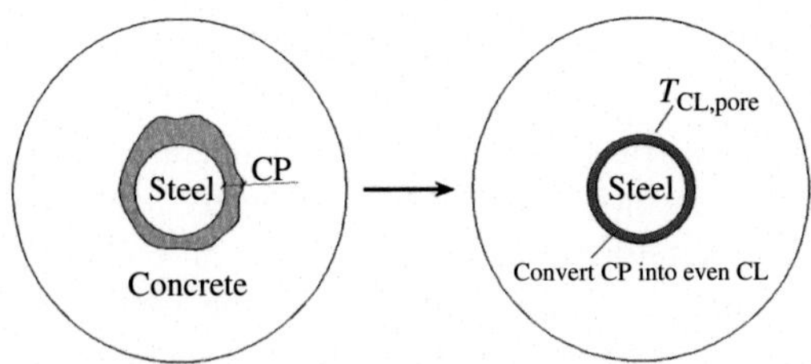

FIGURE 10.3 Conversion from the thickness of CP (T_{CP}) to the thickness of CL ($T_{CL,pore}$).

where ϕ is the porosity of cement mortar around the steel bar, R is the radius of the steel, and θ is the angle between the radius of the integral point and the starting point.

For uniform corrosion, CL is evenly distributed along the steel. The thickness of the CL is assumed to equal $\overline{T_{CL}}$. Then, the average thickness of the filled area, $\overline{T_{CP}}$, can be deduced from Eq. (10.1) as follows:

$$\overline{T_{CP}} = \begin{cases} k_T \times \overline{T_{CL}}, & \overline{T_{CL}} < T_{CL}^{cr} \\ T_{CP}^{max} = k_T \times T_{CL}^{cr}, & \overline{T_{CL}} \geq T_{CL}^{cr} \end{cases} \tag{10.3}$$

According to Eq. (10.2), $T_{CL,pore}$ can be expressed as follows:

$$T_{CL,pore} = \phi \times \overline{T_{CP}} \tag{10.4}$$

Combining Eq. (10.4) with Eq. (10.3), $T_{CL,pore}$ can be expressed as follows:

$$T_{CL,pore} = \begin{cases} \phi \times k_T \times \overline{T_{CL}}, & \overline{T_{CL}} < T_{CL}^{cr} \\ \phi \times T_{CL}^{max} = \phi \times k_T \times T_{CL}^{cr}, & \overline{T_{CL}} \geq T_{CL}^{cr} \end{cases} \tag{10.5}$$

It is well known that the porosity of concrete depends on its quality. According to Eq. (10.5), $T_{CL,pore}$ also depends on the quality of the concrete and the value of k_T is determined from the relation curve of $T_{CP} - T_{CL}$.

As has been stated, CP thickness did not reach the maximum value before the concrete surface cracking; the concrete surface cracking moment should be somewhere in the sideling (oblique) line in Fig. 10.2. Therefore, the first half of the equations $\overline{T_{CP}}$ and $T_{CL,pore}$ when $\overline{T_{CL}} < T_{CL}^{cr}$ in Eqs. (10.3) and (10.5) are used in the concrete surface cracking model.

10.2.3 Nominal Ratio Between the Corrosion Products Volume and the Basic Steel Volume

If the volume of original steel that has corroded is defined as V_{steel}, and the volume of the total rust generated from the basic steel is V_{rust}, then the ratio between the corrosion product volume and the basic steel volume n can be expressed as follows:

$$n = \frac{V_{\text{rust}}}{V_{\text{steel}}} \tag{10.6}$$

where n normally varies from 2 to 4.

In the actual process of concrete cracking caused by steel corrosion, some corrosion products form the CL and the other corrosion products fill the adjacent concrete pores. Therefore, the total volume of corrosion products, V_{rust}, is the sum of the two parts. To distinguish the two types of steel corrosion products (ie, the corrosion products that generates the CL and the corrosion products that forms the CP), the volume of corrosion that generate the CL is $V_{\text{rust,CL}}$, and the volume of corrosion products that form the CP is $V_{\text{rust,CP}}$. The total corrosion volume can thus be expressed as follows:

$$V_{\text{rust}} = V_{\text{rust,CP}} + V_{\text{rust,CL}} \tag{10.7}$$

$V_{\text{rust,CL}}$ is the volume of rust that actually induced the concrete cracking. Therefore, we define the nominal ratio between the corrosion product volume and the basic steel volume, n_0, as follows:

$$n_0 = \frac{V_{\text{rust,CL}}}{V_{\text{steel}}} \tag{10.8}$$

The relationship between n_0 and n is expressed as follows:

$$n_0 = \frac{n}{\alpha_{\text{n}}} \tag{10.9}$$

where α_{n} is the conversion coefficient of the ratio between the corrosion product volume and the basic steel volume, as follows:

$$\alpha_{\text{n}} = \frac{n}{n_0} = \frac{V_{\text{rust}}/V_{\text{steel}}}{V_{\text{rust,CL}}/V_{\text{steel}}} = \frac{V_{\text{rust,CL}} + V_{\text{rust,CP}}}{V_{\text{rust,CL}}} = 1 + \frac{V_{\text{rust,CP}}}{V_{\text{rust,CL}}} \tag{10.10}$$

The rust volume can be approximated by the rust thickness as:

$$\frac{V_{\text{rust,CP}}}{V_{\text{rust,CL}}} \approx \frac{T_{\text{rust,CP}}}{T_{\text{rust,CL}}} = \frac{T_{\text{CL,pore}}}{T_{\text{CL}}} \tag{10.11}$$

Combining Eqs. (10.10) and (10.11), the following relation is obtained:

$$\alpha_{\text{n}} = 1 + \frac{T_{\text{CL,pore}}}{T_{\text{CL}}} \tag{10.12}$$

Eq. (10.10) shows that a larger α_{n} corresponds to a larger $V_{\text{rust,CP}}$, indicating more corrosion products in the concrete pores. If $\alpha_{\text{n}} = 1.0$, then the corrosion filling into the concrete pores in the steel–concrete interface is neglected.

According to Eqs. (10.5) and (10.12), before concrete surface cracking, α_n can be represented as follows:

$$\alpha_n = 1 + \phi \times k_T \tag{10.13}$$

Eq. (10.13) shows that as ϕ increases, α_n also increases, indicating that concrete with larger porosity can accommodate more corrosion products.

Considering the effect of CP on the corrosion-induced concrete surface cracking model, the analysis of the corrosion-induced concrete cracking process is the same as that stated in chapter "Damage Analysis and Cracking Model of Reinforced Concrete Structures With Rebar Corrosion"; simply use the nominal volume expansion rate, n_0, instead of the ratio of the corrosion product volume and the basic steel volume, n.

10.3 TIME FROM CORROSION INITIATION TO CONCRETE SURFACE CRACKING

10.3.1 Faraday's Law

Faraday's second law of electrolysis [4] is that "for a given quantity of DC electricity (electric charge), the mass of an elemental material altered at an electrode is directly proportional to the element's equivalent weight." The equivalent weight of a substance is equal to its molar mass divided by the change in oxidation state that it undergoes on electrolysis and is often equal to its charge or valence.

Faraday's second law gives the relationship between the weight loss of the substance and the electric charge. Based on this Faraday's law, the relationship between the weight loss of a steel bar embedded in concrete, m, and the electric charge, Q, which is decided by the corrosion current running through the steel bar and the time, can be established as:

$$m = \frac{M_{\text{steel}} Q}{zF} \tag{10.14}$$

where m is the mass of steel liberated at an electrode in grams, Q is the total electric charge passed through the steel bar in Coulombs, F is the Faraday constant ($F = 96485$ coulomb/mol), M_{steel} is the molar mass of steel ($M_{\text{steel}} = 55.85$ g/mol), and z is the valence number of the ions of steel. In the case of iron (Fe) and the ferrous (Fe^{2+}), $z = 2$.

For a given length of steel bar, L (in cm), the mass loss of steel bar at concrete surface cracking m_{cr} can be calculated as:

$$m_{cr} = V_{cr,\text{steel}} \times \rho_{\text{steel}} = \pi dL \times \overline{\delta}_{cr} \times 10^{-4} \times \rho_{\text{steel}} \tag{10.15}$$

Where ρ_{steel} is 7.85 g/cm^3, d is the diameter of the steel bar (in cm), and $\overline{\delta}_{cr}$ is the average radial loss of the steel bar at the concrete surface cracking moment in μm, which has the same meaning as δ_{cr}.

The total charge Q is the electric current I_{corr} integrated over time t:

$$Q = \int_0^t I_{corr} d(t) \tag{10.16a}$$

In the simple case of constant-current electrolysis:

$$Q = I_{corr} t \tag{10.16b}$$

For the steel bar with a length of L in cm, the relationship between the corrosion current, I_{corr}(A), and the corrosion current density, i_{corr} (μA/cm^2), is as follows:

$$I_{corr} = \pi d L \times 10^{-6} \times i_{corr} \tag{10.17}$$

Substituting Eq. (10.17) into Eq. (10.16b) yields:

$$Q = \pi d L \times 10^{-6} \times i_{corr} t \tag{10.18}$$

Substituting Eqs. (10.18) and (10.15) into Eq. (10.14) yields:

$$\delta_{cr} \times 10^{-4} \times \rho_{steel} = \frac{M_{steel} Q}{zF} \times 10^{-6} \times i_{corr} t_{cr} \tag{10.19}$$

Therefore, the time to concrete surface cracking can be estimated by:

$$t_{cr} = \frac{zF\delta_{cr} \times 10^2 \times \rho_{steel}}{M_{steel} \times i_{corr}} \tag{10.20}$$

Theoretically, Eq. (10.20) can be used to estimate the time from corrosion initiation to concrete surface cracking as long as the radial loss of steel bar at concrete surface cracking δ_{cr} and the corrosion current density i_{corr} can be obtained.

The previous research work in this area has proposed a number of models to predict δ_{cr}, as stated in chapter "Introduction." The model proposed in this book (see Section 10.2 and chapter: Damage Analysis and Cracking Model of Reinforced Concrete Structures With Rebar Corrosion), which considers the CP at the interface of a steel bar and concrete, is recommended for estimating δ_{cr}.

The model to characterize the corrosion rate, i_{corr}, has not been well developed because of the complexities of the corrosion process in concrete, which is discussed in Section 10.3.2. However, some devices has been developed to measure i_{corr}, such as the GECOR 8 corrosion rate meter and Rapid Corrosion Test (RapiCor), which can make the service behavior prediction of the target RC members more reliable.

10.3.2 Corrosion Rate

As stated in Section 10.3.1, the corrosion current density i_{corr}, which is known as the corrosion rate in several studies, is an important parameter for quantitatively predicting the service life of reinforced concrete structures subject to steel corrosion in concrete.

The corrosion of steel in concrete is highly influenced by factors such as temperature, relative humidity, resistivity of the surrounding concrete, and the supply of oxygen to the cathode site. These factors simultaneously influence the corrosion process and cannot be isolated from each other.

10.3.2.1 Temperature

In general, corrosion rates increase as the temperature increases. A regression model by Liu and Weyers [1] is:

$$i_1 = i_2 e^{2283\left(\frac{1}{T_2}-\frac{1}{T_1}\right)} \tag{10.21}$$

where i_1 is the corrosion current density at temperature T_1, i_2 is the corrosion current density at temperature T_2, T_1 is the temperature of the concrete at measurement (in degrees K), and T_2 is the temperature at which one wants to know the corrosion current density (in degrees K).

Notably, because the changes of temperature in concrete will also result in changes of the other parameters, the overall effect of temperature on the corrosion rate in concrete is highly complex and controlled by interactions of other factors.

In a dry environment, because the surrounding concrete has high resistance, the corrosion rate of the steel bar might be small, even at a high temperature.

10.3.2.2 Humidity

The relative humidity of concrete has a major effect on the corrosion process by influencing the diffusion of oxygen and the ionic resistance of the concrete. The ohmic resistance of concrete may change significantly from more than $10^4\ \Omega$ in a dry environment to approximately several hundred ohms when concrete is saturated [1].

As has been explained in chapter "Steel Corrosion in Concrete," steel corrosion in concrete requires a sufficient supply of oxygen for the cathodic reaction to occur and for the moisture to act as an electrolyte of low resistance. Therefore, two main factors controlling the corrosion rate are the availability of dissolved oxygen surrounding the cathodic areas and the electrical resistance of the concrete surrounding the steel. Generally, the corrosion rate increases as the relative humidity increases when the corrosion is under resistive control, whereas the corrosion rate decreases as the relative humidity of corrosion increases under the control of a cathodic reaction [2].

In fully saturated concrete or very dry concrete, the corrosion rate is notably low.

10.3.2.3 Chloride Content

The experimental work of Liu and Weyers [1] shows that the corrosion rate increases as the amount of chloride content increases in concrete, which indicates that the corrosion rate will be higher at high levels of chloride contamination. This is the result of an increase in the conductivity of concrete as the chloride ions increase, and chloride ions can also combine with ferrous ions to form a water-soluble product that can accelerate the corrosion processes.

10.3.2.4 Bar Diameter

For a given environment, the bar diameter defines the magnitude of the corrosion current and, consequently, of the expansive disruptive corrosion forces [3]. A bigger bar provides lower electrical resistance and, hence, higher corrosion currents. More significantly, a bigger bar diameter generates more corrosion products.

10.3.2.5 Corrosion Time

The corrosion time has an effect on the corrosion rate. The corrosion rate decreases rapidly at the early stage (first year after initiation) and then tends to reach a near-constant value [1]. This is due to a reduction in the anode and cathode area ratios and also results from the formation of the rust products on the steel surface, which slows the diffusion of the iron ions away from the steel surface.

Conversely, the corrosion rate is influenced by factors such as temperature, relative humidity, and chloride content. These factors clearly vary with time. For example, the corrosion rate of a steel bar in summer might be higher than that in winter, and the corrosion rate of a steel bar after rain might be higher than that before rain.

Therefore, the corrosion rate is actually a time-dependent value. Strictly speaking, the assumed simple case of a constant current cannot reflect reality. However, as stated, the model used to characterize the corrosion rate has not been well developed. To further the application of Faraday's law in the prediction of steel corrosion in concrete, it is necessary to develop a model to characterize the corrosion rate, reflect the influence of time, and adjust to an equivalent value according to the structures' exposure conditions. Thus, the service life of reinforced concrete structures can be better predicted.

10.4 DISCUSSION OF NONUNIFORM CORROSION SITUATION

The uniform distribution of corrosion is an ideal state; in practical applications, most of the steel bars in concrete corrode nonuniformly, as stated in chapter "Nonuniform Distribution of Rust Layer Around Steel Bar in Concrete." Including the CP in the concrete surface cracking model induced

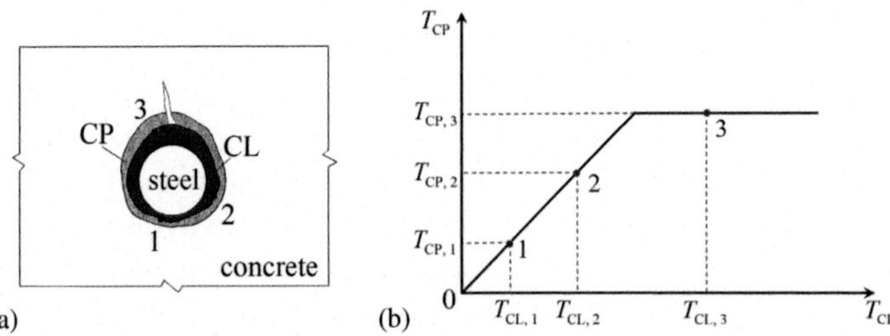

FIGURE 10.4 Nonuniform corrosion layer and the corresponding CP thickness. (a) Nonuniform corrosion layer. (b) Corresponding CP thickness.

by nonuniform corrosion is much more complicated than in cases in which the corrosion can be assumed to be uniform. That complication exists because the degree of corrosion around the perimeter of the steel bar varies; certain regions with serious corrosion enter the platform (shown as point 3 in Fig. 10.4), whereas other regions with less corrosion may still be in the increasing sideling (oblique) line, as illustrated by points 1 and 2.

To calculate the thickness of the CP, T_{CP}, the distribution of the nonuniform CL (ie, the distribution of T_{CL}), which varies around the perimeter of the steel bar, first needs to be known. The geometry-based mathematical model $f(T_{CL})$ has been investigated in chapter "Nonuniform Distribution of Rust Layer Around Steel Bar in Concrete," from which the distribution of the CP thickness, T_{CP}, around the steel perimeter, $f(T_{CP})$, can be established using Eq. (10.1). The conversion CL thickness, $T_{CL,pore}$, can then be obtained by integration using Eq. (10.3).

Theoretically, as long as the mathematical distribution model of the nonuniform CL is given, the $T_{CP}-T_{CL}$ relation model is established, the steel corrosion at the concrete surface cracking induced by nonuniform corrosion can then be predicted. Currently, however, the related research is far from maturity, and more experimental work and theoretical analysis are needed in this area.

10.5 DISCUSSION OF INFLUENCE OF LOADING ON THE CRACKING MODEL

In reality, most RC members are designed for sustaining different types of loads. However, the corrosion-induced concrete cracking process investigated, including the studies in this book, is mainly based on specimens that are not subject to loading. The influence of loading on the cracking model needs to be further studied.

10.5.1 Force Contributed by the Mechanical Interlocking

The deformed bar is normally used for the longitudinal reinforcement to sustain the stress induced by loading. The mechanical interlocking of the deformed steel bar is enhanced by the geometry of the ribs along the length of the steel bar. The force transfer mechanism is due to the mechanical interlocking between the steel ribs and the concrete keys occurring in one of two ways, either in the longitudinal direction to push the concrete keys away (ie, τ_{rib}) or in the radial direction to crash the concrete cover (ie, q_{rib}), as illustrated in Fig. 10.5.

For the corroded RC members subject to loading, the total radial force to crack concrete, q_{total}, should be the sum of the forces from both the steel corrosion and the loading:

$$q_{\text{total}} = q + q_{\text{rib}} \tag{10.22}$$

where q is the radial force induced by steel corrosion (its definition has been illustrated in Fig. 4.1) and q_{rib} is the contribution from the mechanical interlocking between the steel ribs and the concrete keys induced by loading.

It can be noted that most existing models do not consider the influence of loading; however, caution must be exercised when using these models for the prediction of corrosion-induced cracking of RC structures because when the loading is considered, which is the real situation in most cases, the radius loss of the steel bar at the corrosion-induced concrete cracking is less than that calculated from the models that do not consider the loading. Therefore, more research is needed in this area to improve the cracking model.

10.5.2 Intersecting Cracks and Localized Corrosion

For normal reinforced concrete members, the loading-induced narrow cracks are permitted to exist in the concrete cover during their service life as long as the widths of the cracks are within limitations, normally 0.2–0.3 mm in different codes. Typically, these cracks are induced by tensile stress and can be encountered at the bottom of RC beams. Because these cracks traverse in the direction of reinforcement, they are called "intersecting cracks."

In the intersecting cracks regions, easy access is provided for carbon dioxide or chloride, oxygen, and moisture, which may result in localized

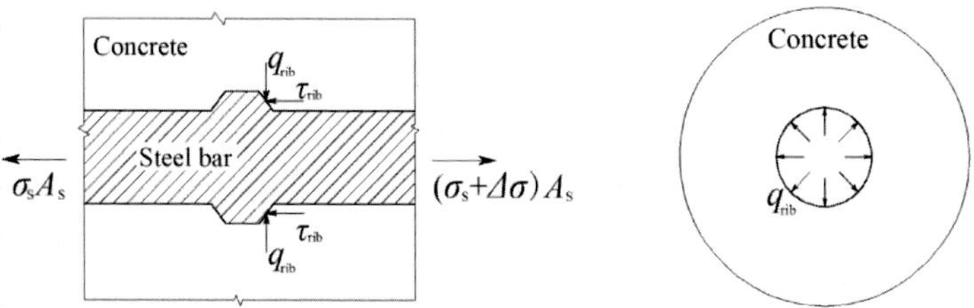

FIGURE 10.5 The mechanical interlocking between the steel ribs and the concrete keys.

corrosion. Further, because of the electrochemical coupling between the localized corrosion spot and the large surrounding passive steel region in the uncracked concrete part, the corrosion macro cell will form. Macro cell coupling is likely to accelerate the corrosion process and leads to a faster rate of radius loss compared to the case of uniform corrosion. Therefore, the areas around intersecting cracks might have a different cracking mode compared with non-cracking area. However, the effect of localized corrosion on the corrosion-induced model is not yet addressed in detail in the literature and needs to be investigated further.

10.6 CONCLUSIONS

1. A $T_{CP}-T_{CL}$ model, which describes the relationship between the thickness of CP and that of CL, was proposed. The parameter k_T in this model was used to describe that the corrosion products filling in the concrete pores and accumulating at the steel–concrete interface occur synchronously.
2. The nominal ratio between the corrosion product volume and the basic steel volume, n_0, was introduced to include the corrosion products in the CP layer. α_n and the ratio between n_0 and n were defined as conversion coefficients to indicate the capacity of concrete to accommodate corrosion products.
3. An improved corrosion-induced concrete cracking model was proposed. In the model, CL accumulation and corrosion products filling occur simultaneously in concrete.
4. The time from corrosion initiation to concrete surface cracking are discussed. Theoretically, this time can be estimated as long as the radial loss of steel bar at concrete surface cracking and the corrosion rate can be obtained. The corrosion rate, which is highly influenced by many factors, is a time-dependent value. The model to characterize the corrosion rate needs to be further developed.
5. More research is needed to improve certain areas of the corrosion-induced cracking model, including the nonuniform corrosion situation and the influence of loading on corrosion-induced cracking process.

REFERENCES

[1] Liu T, Weyers RW. Modeling the dynamic corrosion process in chloride contaminated concrete structures. Cement Concrete Res 1998;28(3):365–79.

[2] Bo Y, Yang LF, Ming W, Bing Li. Practical model for predicting corrosion rate of steel reinforcement in concrete structures. Constr Building Mater 2014;54:385–401.

[3] Rasheeduzzafar Al-Saadoun SS, Al-Gahtani AS. Corrosion cracking in relation to bar diameter, cover, and concrete quality. ASCE, J Mater Civil Eng 2014;4(4):327–42.

[4] <https://en.wikipedia.org/wiki/Faraday%27s_laws_of_electrolysis>.

Notations

A	Area of flaky rust sample
A_r	Total area of rust layer
A_{ri}	Area of the ith part of the rust layer that is separated by the cracks
A_s	Cross-section area of uncorroded steel bar
a_1	Crack width variation coefficient
a_2	Crack width coefficient at the surface of the steel bar
b	Radius of cylinder-reinforced concrete specimens
C	Thickness of concrete cover
c	Cohesive strength of concrete
C_N	Normal concrete
C_P	Very porous concrete
C_S	Very solid concrete with extremely low porosity
CP	Corrosion products-filled paste
CL	Corrosion layer
D	Damage variable
d	Diameter of steel bar
d_ρ	Residual diameter of the steel bar after corrosion
$d_{\rho c}$	Critical diameter of steel at the moment of concrete surface cracking
d_1	Nominal diameter of the steel bar with free-expansion corrosion products
E_c	Elastic modulus of concrete
E'_c	Deformation modulus at the maximum tensile stress
$E_{c,r}$	Elastic modulus of the concrete in the radial direction
$E_{c,\theta}$	Elastic modulus of the concrete in the hoop direction
E_r	Elastic modulus of corrosion products
$E_{r,o}$	Elastic modulus of flaky rust sample
F	Faraday constant
f_c	Compressive strength of concrete
f_t	Ultimate tensile strength of concrete
I_{corr}	Corrosion current of steel bars
k_T	Ratio between T_{CP}^{max} and T_{CL}^{cr}
i_{corr}	Corrosion current density of steel bars
L	Length of steel bar
l_{ci}	Distance from the radius R_i to the surface of steel bar
m	Mass of steel liberated at an electrode
m_{cr}	Mass loss of steel bar at concrete surface cracking
MS	Mill scale
M_{steel}	Molar mass of steel
n	Expansion coefficient of rust, $n = V_{rust}/V_{steel}$

n_0	Nominal ratio between the corrosion product volume and the basic steel volume, $n_0 = V_{rust,CL}/V_{steel}$
Q	Total electric charge passed through the steel bar in Coulombs
q	Expansive pressure induced by steel corrosion at steel/concrete interface
q_{max}	Maximum expansive pressure at the steel–concrete interface
q_{Rc}	Radius expansive pressure at the interface between intact and cracked concrete cylinders
q_{rib}	Force between the steel rib and the concrete key in radial direction
R	Radius of steel bar
R_1	Nominal radius of steel bar with free-expansion corrosion products
R_c	Radius at interface between intact and cracked concrete cylinders
ΔR	Thickness of each concrete ring
r	Radial coordinate of cylinder concrete model
T_{CL}	Thickness of corrosion layer
$T_{CL,pore}$	Corrosion layer thickness that is conversed from CP
T_{CL}^{cr}	Critical thickness of corrosion layer corresponding to T_{CP}^{max}
T_{CL}^{inner}	Thickness of corrosion layer at inner surface cracking
$T_{CL}^{surface}$	Thickness of corrosion layer at outer concrete surface cracking
T_{CP}	Thickness of corrosion products-filled paste
T_{CP}^{max}	Maximum of the thickness of corrosion products-filled paste
T_m	Thickness of mill scale
T_r	Thickness of rust layer
$T_{r,max}$	Maximum thickness of rust layer
$T_{r,min}$	Minimum thickness of rust layer
$\overline{T_{CL}}$	Average thickness of corrosion layer
$\overline{T_{CP}}$	Average thickness of corrosion products-filled paste
t_{cr}	Corrosion-induced surface cracking time
u_r	Radial deformation of cylinder concrete model
$V_{cr,steel}$	Volume loss of steel bar at concrete surface cracking
V_{rust}	Total rust generated from basic steel
$V_{rust,CL}$	Volume of corrosion that generates CL, which induces the concrete cracking
$V_{rust,CP}$	Volume of corrosion that generates CP
V_{steel}	Volume of original steel
W_c	Critical crack width at the moment of concrete surface cracking
W_i	Total circumferential crack width of at radius R_i
W_s	Total crack width at concrete surface
w_{ki}	Circumferential crack width of a single crack at radius R_i
w_{sk}	Crack width of a single crack at the edge of a slice
z	Number of the ions of steel
α_n	Conversion coefficient of the ratio between n and n_0, $\alpha_n = n/n_0$
γ	Reflecting the width variation of corrosion-induced crack, $\gamma = \arctan a_1$
Δ	Replacement percentage of recycled coarse aggregate
δ	Steel radial loss induced by corrosion
δ_c	Deformation of concrete at the steel–concrete interface
δ_{cr}	Critical radial loss of steel bar at concrete surface cracking
δ_{crack}	Radial loss of steel bar during concrete cracking period; the corrosion products fill the space within the cracks

δ_{pore}	Radial loss of steel bar during free-expansion period; the corrosion products fill the porous voids around the steel–concrete interface
δ_r	Deformation of corrosion products at the concrete–rust interface
δ_{stress}	Radial loss of steel bar produced between stress initiation in concrete cover and concrete surface cracking
δ_{stress}^{inner}	Radial loss of steel bar during concrete inner surface cracking
$\delta_{stress}^{surface}$	Radial loss of steel bar during concrete outer surface cracking
ε_r	Radial strain induced by steel corrosion in concrete cover
ε_r^e	Radial elastic strains induced by steel corrosion in concrete cover
ε_t	Ultimate tensile strain of concrete, $\varepsilon_t = f_t/E_c$
ε_θ	Hoop strain induced by steel corrosion in concrete cover
ε_θ^e	Hoop elastic strains induced by steel corrosion in concrete cover
θ	Polar coordinate of cylinder model
λ_1	Nonuniform coefficient of rust layer
λ_2	Spread coefficient of rust layer
λ_3	Uniform coefficient of rust layer
λ_4	Fitting parameter
ν_c	Poisson ratio of concrete
ν_r	Poisson ratio of corrosion products
ξ	Height of flaky rust sample
ρ	Steel corrosion
ρ_{cr}	Critical steel corrosion at concrete surface cracking
σ_r	Radial stress induced by steel corrosion in concrete cover
σ_s	Stress in steel bar
σ_θ	Hoop stress induced by steel corrosion in concrete cover
τ_{rib}	Force between the steel rib and the concrete key in longitudinal direction
φ	Internal friction angle of concrete
ψ	Height-to-area ratio of flaky rust sample, $\psi = \xi/\sqrt{A}$
ϕ	Porosity of concrete

Index

Note: Page numbers followed by "*f*" and "*t*" refer to figures and tables, respectively.

D

R

S

Printed in the United States
By Bookmasters